# Palgrave Studies in Agricultural Economics and Food Policy

**Series Editor**
Christopher Barrett
Cornell University
Ithaca, NY, USA

Agricultural and food policy lies at the heart of many pressing societal issues today and economic analysis occupies a privileged place in contemporary policy debates. The global food price crises of 2008 and 2010 underscored the mounting challenge of meeting rapidly increasing food demand in the face of increasingly scarce land and water resources. The twin scourges of poverty and hunger quickly resurfaced as high-level policy concerns, partly because of food price riots and mounting insurgencies fomented by contestation over rural resources. Meanwhile, agriculture's heavy footprint on natural resources motivates heated environmental debates about climate change, water and land use, biodiversity conservation and chemical pollution. Agricultural technological change, especially associated with the introduction of genetically modified organisms, also introduces unprecedented questions surrounding intellectual property rights and consumer preferences regarding credence (i.e., unobservable by consumers) characteristics. Similar new agricultural commodity consumer behavior issues have emerged around issues such as local foods, organic agriculture and fair trade, even motivating broader social movements. Public health issues related to obesity, food safety, and zoonotic diseases such as avian or swine flu also have roots deep in agricultural and food policy. And agriculture has become inextricably linked to energy policy through biofuels production. Meanwhile, the agricultural and food economy is changing rapidly throughout the world, marked by continued consolidation at both farm production and retail distribution levels, elongating value chains, expanding international trade, and growing reliance on immigrant labor and information and communications technologies. In summary, a vast range of topics of widespread popular and scholarly interest revolve around agricultural and food policy and economics. The extensive list of prospective authors, titles and topics offers a partial, illustrative listing. Thus a series of topical volumes, featuring cutting-edge economic analysis by leading scholars has considerable prospect for both attracting attention and garnering sales. This series will feature leading global experts writing accessible summaries of the best current economics and related research on topics of widespread interest to both scholarly and lay audiences.

More information about this series at
http://www.palgrave.com/gp/series/14651

John M. Antle • Srabashi Ray

# Sustainable Agricultural Development

## An Economic Perspective

palgrave
macmillan

John M. Antle
Department of Applied Economics
Oregon State University
Corvallis, OR, USA

Srabashi Ray
Department of Applied Economics
Oregon State University
Corvallis, OR, USA

ISSN 2662-3889          ISSN 2662-3897   (electronic)
Palgrave Studies in Agricultural Economics and Food Policy
ISBN 978-3-030-34598-3          ISBN 978-3-030-34599-0   (eBook)
https://doi.org/10.1007/978-3-030-34599-0

Cover illustration: © fotoVoyager

This Palgrave Macmillan imprint is published by the registered company Springer Nature Switzerland AG.
The registered company address is: Gewerbestrasse 11, 6330 Cham, Switzerland

# Foreword

Modern worries about environmental degradation focus appropriately on anthropogenic causes, that is, on the human behaviors that despoil air, climate, land, and water and imperil nonhuman species. Agriculture is among the human activities that most intimately links our behaviors to natural systems, with bidirectional feedback since the state of natural systems heavily influences agricultural productivity just as agricultural practices regulate the ecosystems in which they take place. Ever since humankind began domesticating plants and animals for agricultural production several millennia ago, our species has been actively changing natural ecosystems, altering them to suit our consumptive and aesthetic purposes. But climate, water quality, biodiversity loss, deforestation, and other environmental crises have emerged due to our imperfect understanding of the complex feedback intrinsic to agroecosystems, spillover effects that make it easy for decision-makers to ignore the full range of consequences of our actions, and present biased preferences. So today we face multiple environmental crises, each directly linked to agriculture.

The world has confronted agriculture-related crises before; we can learn from past experience. For example, roughly fifty years ago, Malthusian concerns about rapid human population growth, a global food price spike, and recurring famines in the late 1960s and early 1970s led to a world food crisis. This episode rapidly concentrated policymakers' and researchers' attention on global hunger. Major investments followed, focusing on rapidly expanding the supply of staple cereals, roots, and tubers that comprise the base diets of the poor worldwide, and launching and expanding social protection policies with food assistance programs as a centerpiece. The resulting Green Revolution, global early warning systems, and dramatic expansion of cash and in-kind transfer programs successfully drove down real food prices and virtually eliminated famines for several decades. The core objective of averting mass hunger was accomplished. For two generations now, the world has been providing adequate calories for roughly ninety million additional people year after year, a remarkable accomplishment.

But success in addressing the world food crisis of the 1970s by targeting global hunger came with a heavy environmental price, on which world leaders are now appropriately concentrating, manifest in the sustainable development goals and related declarations. Moreover, environmental degradation now threatens the agricultural productivity and nutritional gains of recent decades. Extreme climate events and soil and water degradation have contributed to modest upticks in the number of undernourished people globally the past three years. It is (past) time to shift objectives: to safeguard the planet while maintaining agricultural productivity growth on which economic development fundamentally depends; hence, the importance of sustainable agriculture, and of this book.

In this impressive volume, John Antle and Srabashi Ray lucidly explain the intricate interrelationships between agriculture and the environment, the path followed over the past half century as the world focused almost single-mindedly on growing global per capita calorie supplies, the environmental impacts of modern agricultural practices, and the current state of the debate on sustainable agriculture. Most importantly, they carefully lay out the tradeoffs intrinsic to the choices faced by farmers, agribusinesses, government policymakers, and individual consumers. They also flag where synergies might arise, where economic and environmental advance might be mutually reinforcing through agricultural development.

Ultimately, Antle and Ray offer a hopeful analysis. We have faced and overcome agriculture-related crises before and can do so again. But this is not an easy task. It will take political will, but it will equally require outstanding science. Some technological and institutional changes can foster total factor productivity growth that enables simultaneous improvements in agricultural, economic, environmental, and social indicators. But more generally, we face tradeoffs that must be assessed carefully and honestly. This requires marshaling evidence and insights from multiple disciplines, a task for which few, if any, scientists are better equipped than John Antle. Careful analysis and accurate data are essential to evaluate these tradeoffs, to find the feasible sustainable agricultural development paths, and avoid the dangers of either continued environmental degradation or increased unnecessary undernutrition. For a generation, Professor Antle has been advancing the research frontier in developing our analytical tool kit for tradeoffs analysis. His work with Ray and other coauthors uniquely equips these authors to guide the reader through the myriad issues that link ever-changing agricultural practices to economic and environmental outcomes of global importance. Indeed, they are the ideal intellectual guides for those of us interested in this topic.

The powerful insights Antle and Ray offer in this volume are too numerous to summarize in this brief foreword. In refreshingly clear prose, this outstanding volume lays out the central issues in accessible terms and compactly summarizes a deep and complex literature with remarkable precision and rigor. Serious students of sustainable agriculture need to read this volume.

It is a great pleasure to include John Antle and Srabashi Ray's excellent volume in the Palgrave Studies in Agricultural Economics and Food Policy series. I learned a good deal by reading it and expect many others will as well. This book should prove an essential reference to anyone striving to understand the origins and evolution of agricultural and food policy in modern society.

Cornell University, Ithaca, NY, USA                    Christopher B. Barrett

# Preface

Sustainability is now widely accepted as a guiding concept and goal for our economies and societies, and for agriculture and the food system. But it is also widely accepted that modern societies, as well as agriculture and the food system, are now on largely unsustainable trajectories. We have entered the Anthropocene, the era in which humans are the dominant force affecting the state of Earth systems on which life depends. One of the grand challenges facing humanity is to put itself on a sustainable trajectory.

We conclude this book by saying that there is reason for cautious optimism that we can meet this challenge in the coming decades. Fifty years ago, many people justifiably doubted the ability of the world to feed itself, and 'limits to growth' were predicted as humans exhausted natural resources such as oil. Despite a rapidly growing world population, those dire prospects were avoided through applications of science and technology as well as institutional and policy innovations. Humans created a remarkable global food system that allows more people than ever before to have more than adequate diets and attain the longest life expectancy in history. Much of this progress was possible because humans developed ways to discover and use more fossil fuels than once was thought possible.

But these very successes in agricultural development, and in fossil-fueled economic development, have created huge challenges in environmental and social dimensions that were largely unanticipated until recently. Just as the abundance of fossil fuels turns out to be the largest obstacle to a stable and livable climate, the indiscriminate pursuit of cheap calories has led agriculture and the food system down an unsustainable pathway.

The good news is that we now have many tools that can be used to design and build more sustainable agricultural systems, and we introduce the reader to some of them in this book. Using these tools, researchers, together with stakeholders, are making progress to identify the changes in household and producer behavior that could limit climate change to 1.5 °C, and they are moving agriculture and the food system in more sustainable directions. First and foremost, we know that to limit climate change, the global energy system must move rapidly away from reliance on fossil fuels. But also, agriculture and the food system must change—for example, recent studies show that substantially reducing meat consumption could have a large impact on greenhouse gas emissions and have other environmental and health benefits. At the same time, meat production, consumption, and trade are growing rapidly, particularly in the regions of the world undergoing the transition from low- to middle-income status. And, many people argue that increasing meat consumption can play a valuable role in improving the nutrition in the poor regions of the world. What is lacking in the debate over sustainable development, in our view, is to go beyond the identification of goals and the design of possible sustainable development pathways, to implement feasible actions—technologies and policies—that will move today's agricultural systems in more sustainable directions.

In this book, we provide an explanation—from the perspective of economics—for this gap between goals and the state of the world. A central theme of this book is that agricultural systems are complex, diverse, and heterogeneous. These characteristics mean that there are various synergies and tradeoffs as we attempt to improve the performance of agricultural systems. And for the most part, attempts to change them focus on particular aspects of their performance. Yet, as we show in this book, there are multiple important factors in each of the three dimensions of sustainability—economic, environmental, and social. This means that there are rarely simple, one-size-fits-all solutions, and there will inevitably be winners and losers as the systems change. The example of the recent expansion of deforestation in the Amazon driven by economic and policy changes in China and the United States is but one example of this complexity. Addressing such challenges requires an understanding of the systems, as well as an effective participatory process to envision and design feasible pathways. It also requires implementation strategies that mobilize consumers, food producers, and all of civil society to work toward solutions.

Corvallis, OR                                                        John M. Antle
                                                                    Srabashi Ray

# Acknowledgments

We'd like to thank Chris Barrett for the opportunity to contribute to this book series, and Sophia Siegler for keeping us on a sustainable pathway toward a successful book. Over the years, John Antle has benefitted from many productive collaborations around the topic of this book. Srabashi Ray is grateful to peers, mentors, friends, and family for their support and encouragement in her professional endeavors. Thanks to all.

# Praise for *Sustainable Agricultural Development*

"This important book clearly lays out the complex challenges facing agricultural development at local-to-global scales. Despite the great success of the current food system, it faces two major issues: climate change and malnutrition. Through their economic lens, the authors explicate the participatory processes and analytical tools needed to find solutions to these challenges. They offer pathways to sustainable agricultural development for both developing and high-income countries, focusing on methods by which to create them. The Agricultural Model Intercomparison and Improvement Project (AgMIP) welcomes this insightful book, and will utilize its methods in major assessments of climate change and the global food system."

—Cynthia Rosenzweig, *Senior Research Scientist, NASA Goddard Institute for Space Studies and the Earth Institute*

"Antle and Ray provide a comprehensive assessment of the synergies and trade-offs in achieving sustainable development with a much needed focus on food and agricultural systems. This book provides timely input to the current debates and controversies around the state and future of global food systems. The authors' cautious optimism that we will eventually get it right and will embark on a path of sustainable development at the local and global scales is appealing. A must read for all of us interested in a future world that can sustainably feed nine billion people."

—Prabhu Pingali, *Professor of Applied Economics, Cornell University*

"Finding pathways to sustainable development is the challenge of our time. Antle and Ray provide a readable, non-technical account of the central role that agriculture must play in meeting that challenge. Using examples from developing and industrialized regions, they explain the concepts and analytics that are used by scientists and economists to design more sustainable agricultural development pathways. Essential reading for anyone who wants to understand agriculture's role in sustainable development."

—Pramod Aggarwal, *South Asia Research Program Director, Research Program on Climate Change, Agriculture and Food Security*

# Contents

# Abbreviations

| | |
|---|---|
| AgMIP | Agricultural Model Inter-comparison and Improvement Project |
| BCA | Benefit-Cost Analysis |
| BMI | Body Mass Index |
| CAADP | Comprehensive Africa Agriculture Development Program |
| CFC | Chlorinated Flurocarbons |
| CGIAR | Consultative Group for International Agricultural Research |
| ERS | Economic Research Service |
| EU | European Union |
| FABLE | Food, Agriculture, Biodiversity, Land, and Energy |
| FAIR | Findable, Accessible, Interoperable, and Reusable |
| FAO | Food and Agriculture Organization |
| FIES | Food Insecurity Experience Scale |
| GAEZ | Global Agro-Ecological Zone |
| GDP | Gross Domestic Product |
| GHG | Green House Gas |
| GMO | Genetically Modified Organism |
| GNI | Gross National Income |
| GR | Green Revolution |
| GYGA | Global Yield Gap Atlas |
| HDDS | Household Dietary Diversity Score |
| HFSSM | Household Food Insecurity Access Scale |
| IBFS | Income-Based Food Security |
| ICRISAT | International Crops Research Institute for the Semi-Arid Tropics |

| | |
|---|---|
| IFPRI | International Food Policy Research Institute |
| IGP | Indo-Gangetic Plain |
| IPCC | Intergovernmental Panel on Climate Change |
| IPL | International Poverty Line |
| IPM | Integrated Pest Management |
| MOSPI | Ministry of Statistics and Programme Implementation |
| NBS | National Bureau of Statistics |
| NCP | North China Plain |
| NECP | Northeast China Plain |
| OECD | Organisation for Economic Co-operation and Development |
| PRD | Pearl River Delta |
| RCP | Representative Concentration Pathway |
| RCT | Randomized Controlled Trial |
| RIA | Regional Integrated Assessment |
| SB | Sichuan Basin |
| SDG | Sustainable Development Goal |
| SSP | Shared Socioeconomic Pathway |
| TOA-MD | Tradeoff Analysis Model for Multi-Dimensional Impact Assessment |
| UN | United Nations |
| US | United States |
| WEAI | Women's Empowerment in Agriculture Index |
| YP | Yangtze Plain |

# List of Figures

# List of Tables

# List of Boxes

# 1

# Introduction

The first agricultural revolution that took place ten millennia ago among a global population of a few million Homo sapiens made possible a transformation from small groups of hunter-gatherers to settled, increasingly urbanized societies supported by farmers—people who specialized in growing domesticated crops and livestock. As humanity approaches a population of 10 billion this century—with the majority living in cities and enjoying an unprecedented material standard of living—it is becoming clear that the phenomenal global economic growth that made this stunning success of Homo sapiens possible may not be sustainable. Most emblematic of the human condition is the dependence of the global economy on the fossil fuels that made this unprecedented human and economic growth possible, but that also drives global climate change and other environmental changes that threaten its sustainability.

Even with the technological advances that have made modern economic growth possible, human survival will continue to depend on its first industry, agriculture. Indeed, the unprecedented growth in human population that has occurred over the past century would not have been possible without the growth in food production made possible by second agricultural revolution—often called the Green Revolution of the

© The Author(s) 2020
J. M. Antle, S. Ray, *Sustainable Agricultural Development*, Palgrave Studies in Agricultural Economics and Food Policy, https://doi.org/10.1007/978-3-030-34599-0_1

Twentieth Century. But people want not just to survive but to realize their potential in the many dimensions of life that they value. Sustainable development is the concept that is now being widely used to embody this concept of economic and social change that improves the well-being of current as well as future generations within the limits of the natural world. The premise underlying this book is that to achieve this vision of sustainable development, agriculture also must change and evolve in ways compatible with these human goals. This is what we mean by sustainable agricultural development.

Our goal in this book is to help the reader understand the key role agriculture must play in sustainable development, by providing an economic perspective on how we can understand, evaluate, and improve the sustainability of agriculture around the world. We think an economic perspective on this admittedly multidimensional challenge is valuable for several reasons. The concept of sustainable development is an outgrowth of the global economic growth and development that the world has experienced over the past two centuries. Agricultural development in turn plays a key role in economic growth and social development, operating through the 'demand side' and the 'supply side' of markets for productive resources and agricultural commodities. There is no question that unprecedented advances on the supply side through improvements in agricultural technologies and food systems have been necessary to support the rapid growth in human populations. Yet, in a market-based, global food system, achieving sustainable agricultural development will also depend importantly on the 'demand side,' that is, on the choices and actions of people who live and work outside of agriculture. Increasingly, consumers and citizens will play an important role in the future of agriculture and food systems. They do this by influencing where and how food is produced and how the global food system evolves—what we refer to in this book as the agricultural development pathway.

It is now clear that the development pathway the world is on is not sustainable in many dimensions, including agriculture. We discuss the major agricultural production systems in the world and their characteristics and elaborate on their current status and prospects for sustainability. But it is not difficult to justify the view that the current trajectory of most major agricultural systems is not sustainable. As just one of many examples we discuss in this book, consider the US Midwest agricultural system that is a major contributor to nutrient pollution causing toxic algal blooms in the Great Lakes and the hypoxic 'dead zone' in the Gulf of Mexico. It is also clear that the food system is not ensuring good nutrition and health outcomes for many people in the world. Recent data show that more than half a billion people are undernourished while more than 2 billion are overweight or obese.

The widespread recognition that the world is approaching and in some cases has already crossed critical local, regional, or global ecosystem thresholds or boundaries has led many people to conclude that the goal of economic development must be broadened to sustainable development if continued human progress is to be achieved. In this book, we follow Sachs (2015) in recognizing the twenty-first century as the dawning of the 'Age of Sustainable Development.' We adopt Sachs' view of sustainable development as both an analytical framework in which to understand the processes and consequences of economic and social development and as a normative framework in which to evaluate and set development goals, such as the sustainable development goals (SDG) established by the United Nations Conference on Environment and Development in 2015.

Sachs provides a valuable introduction to agriculture's role in sustainable development. Yet, agriculture's role in sustainable development deserves further elaboration in an analytical and empirical framework consistent with economic as well as ecological and social science principles. Additionally, the options and actions needed to achieve sustainable agricultural development deserve further discussion and elaboration in a way that is accessible outside the scientific literature. Our goal in this book is to provide a nontechnical, accessible primer on sustainable agricultural development and its relationship to sustainable development.

Our approach is based on three analytical pillars. The first pillar is to understand agriculture's role in economic growth, economic development, and sustainable development, interacting through markets and policies with other sectors of economies at local, national, and global scales. The second pillar is to understand agriculture as a highly diverse array of complex physical-biological-human systems—one can say, as managed agroecosystems. Figure 1.1 portrays the food system this way, as

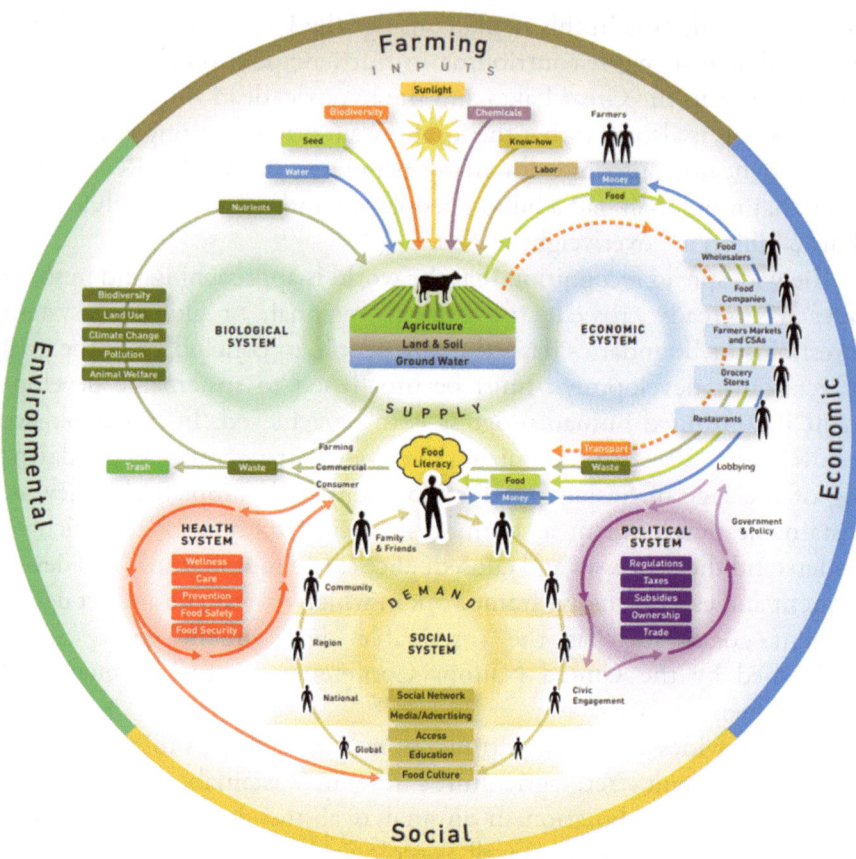

**Fig. 1.1** Food systems map. (Source: WorldLink. Nourish Food System Map, www.nourishlife.org. Copyright WorldLink, all rights reserved)

a set of interconnected subsystems that includes the supply side with farms and agriculture linked through markets to consumers on the demand side, as well as agriculture's interactions with the physical and biological systems. Other key interactions are human health that connects both to food consumption and the environment, as well as governance and policy that cut across all the components of the food system in various ways. The third pillar is the economic perspective of the tradeoffs and synergies among the economic, environmental, and social dimensions of agricultural systems at farm, regional, and global scales. In this book, we describe and use the tradeoff analysis approach to understand sustainable agricultural development. This approach begins with the identification of key indicators of system performance that stakeholders view as relevant to sustainability. Using appropriate data and analytical tools, alternative development pathways are evaluated in terms of their implications for the performance of agricultural systems at local, regional, and global scales.

Chapter 2 provides a 'macro' or big picture view of the role that agriculture plays in economic development and sustainable development, describes some of the aggregate indicators used to set sustainable development goals, and introduces tradeoff analysis as a way to evaluate development pathways. We begin this chapter with the conventional view of economic growth as the driver of economic and social development, and how the increasing human dominance of global ecosystems and subsequent environmental social challenges motivated the concept of sustainable development. We discuss some of the ways that economists have broadened the conventional development model to incorporate sustainability. Next we discuss how sustainability has been used as a normative framework using the sustainable development goals. Finally, we introduce tradeoff analysis as a way to combine the positive and normative approaches to sustainable development into a unified framework to set goals and evaluate progress toward them.

In Chap. 3, we focus on agricultural systems which we describe as a diverse array of production systems that are composed of interconnected physical, biological, and human components. Our approach to agricultural systems involves an understanding of systems at the farm level as well as larger scales. Two key features of agricultural systems are diversity, that is, the different types of systems, and their heterogeneity, that is, the variation in the physical, biological, and human components within each type of system. We introduce the array of economic, environmental, and social indicators that are used to evaluate the diverse and complex agricultural systems throughout the world, and we introduce two examples that illustrate their diversity and complexity. Building on Chap. 2, we also elaborate on the economic rationale for the tradeoff analysis approach to agricultural system sustainability. Finally, we discuss the tools—the computer simulation models and data—used to implement the analysis of agricultural system sustainability, using climate impact assessment as an example.

Chapters 4 and 5 address the challenges of sustainable agriculture in developing countries and industrialized countries. Low-income and transitional countries span a large part of the arable land area of the world. In Chap. 4, we examine the diverse and heterogeneous systems around the world, focusing primarily on the three regions where the vast majority of poor farmers are located: sub-Saharan Africa, South Asia, and East Asia. We discuss the main characteristics of the agricultural systems in these regions, their current status in terms of key sustainability indicators, and some of the challenges they face in moving toward more sustainable development pathways. The challenges of sustainable development in the countries with agricultural sectors that have characteristics of 'industrial' agriculture are quite distinct from those of the developing regions. In these countries, the majority of food is produced on large commercial farms. Farm household incomes are relatively high and often equal or exceed incomes in the non-agricultural sectors for comparable skills and experience. A major economic issue is the financial viability of smaller farms and the ongoing consolidation of land into larger farms. Environmental concerns dominate the sustainability challenges of these systems, due to their reliance on chemical inputs, impacts on water quality and quantity, use of hormones and antibiotics in livestock production, and greenhouse gas emissions. Health and safety of farm workers, rural community viability, and animal welfare are important social challenges.

Chapter 6 addresses the ways that we can, individually and collectively, move agricultural development in more sustainable directions. There is a widespread recognition that the global food system must change to support the goals of sustainable development. Yet, there has been relatively little discussion of how those changes can be implemented. In this chapter, we describe the design and implementation of more sustainable development pathways based on two components. Participatory processes involving scientists and stakeholders are being used to identify sustainability indicators and set goals, and science-based tools are then used to evaluate system performance along the envisioned development pathways. Implementation requires incentives and other policy mechanisms to encourage more sustainable technologies. The complexity of agricultural systems means there is no one-size-fits-all solution to sustainable agricultural development. In Chap. 6, we illustrate the challenges to sustainable development with examples discussed in previous chapters.

The challenge of moving agricultural systems toward more sustainable development pathways is daunting, perhaps even more than the earlier challenge of providing enough calories for a growing global population. The challenge comes from the complexity of agricultural systems—what we describe as their diversity and heterogeneity—and the multiple and often conflicting objectives that society has for agriculture and the food system. The consequence of this complexity and the resulting tradeoffs among desired outcomes is the high degree of difficulty we face in coordinating policies at local, national, and global levels to achieve sustainable development pathways. There is no one simple solution that will achieve a more sustainable development pathway for the diverse agricultural systems across the developing and industrialized countries. So what are we to do? Our answer is: to improve the data and analytical tools we have, so that we can anticipate and identify emerging positive or adverse outcomes based on solid science; to build awareness of the need for change; and to use all levels of action, from individual and local to national and global, to move agricultural systems away from particularly adverse and irreversible outcomes and toward desirable outcomes. We conclude with the good news that, increasingly, we have the data and analytical tools to address the challenges of sustainable agricultural development. Now we must use them.

Finally, we think it is important for the reader to know how we think—one might say, our assumptions and biases. We have already noted the economic perspective—and no doubt some will say biases—that we bring to this book. So to be clear, here are some of the views that guide our thinking about sustainable agricultural development. We see markets as the most important and pervasive institution determining how resources are allocated in local, regional, national, and global context, everywhere in the world—from remote villages to megacities. Adequately functioning legal and political institutions are a prerequisite for a well-functioning market-based economy. Consequently, it is not a coincidence that the poorest countries in the world have extremely weak legal and political institutions. Various factors, including monopoly power and production or consumption externalities, can cause markets to 'fail,' that is, to result in inefficient, socially undesirable, and unsustainable outcomes. The conventional solution to these market failures is government intervention. But we are not naïve about government or policy, because 'government failure' and 'policy failure' are also major problems. Policy can suffer from poor design and implementation and can be captured by interest groups and manipulated in ways that also can lead to socially undesirable and unsustainable outcomes. These policy outcomes can be worse than those resulting from unregulated markets, particularly in places where political and legal institutions are weak and dysfunctional.

We see the goal of sustainable development as enabling people to pursue their own interests as well as collective or social interests. But because what one person does can impact others (there are externalities), this does not mean people are free to pursue their own interests at the expense of other people in the present or in the future. Thus, some constraints on individual freedom through collective action are necessary to achieve sustainable development. In many respects, the crux of the policy debate is how to balance this tradeoff between individual freedom and collective well-being. That being said, we think people everywhere prefer more freedom to less, more income to less, safety to danger, health to sickness, longer to shorter life expectancy, and social harmony to discord.

# Reference

Sachs, Jeffrey D. 2015. *The Age of Sustainable Development.* New York and Chichester, West Sussex: Columbia University Press.

# 2

# Economic Development, Sustainable Development, and Agriculture

## 2.1    Introduction

Our aim in this chapter is to provide a 'big picture' or aggregate view of economic growth, economic development, and sustainable development. Agriculture plays a special role in economic development because it was the 'first industry' of the agricultural revolution that took place about 10,000 B.C. as humans underwent the transition from hunter-gatherers to settled communities. Remarkably, from then until the industrial revolution in the eighteenth century, agricultural productivity was low, and most people in the world had to be farmers even to maintain the relatively small human populations of the preindustrial era, with a global population estimated to be about 1.6 billion in 1800. Because most people and other productive resources were employed in agriculture, the industrial revolution necessarily meant that productive resources—at that time, primarily labor and land, had to be transferred from agriculture to exploit natural resources such as coal and iron that were essential to industrialization. That process began first in England in the eighteenth century, spread throughout Europe and North America and continues today in the parts of the world we now refer to as the 'developing

© The Author(s) 2020
J. M. Antle, S. Ray, *Sustainable Agricultural Development*, Palgrave Studies in Agricultural Economics and Food Policy, https://doi.org/10.1007/978-3-030-34599-0_2

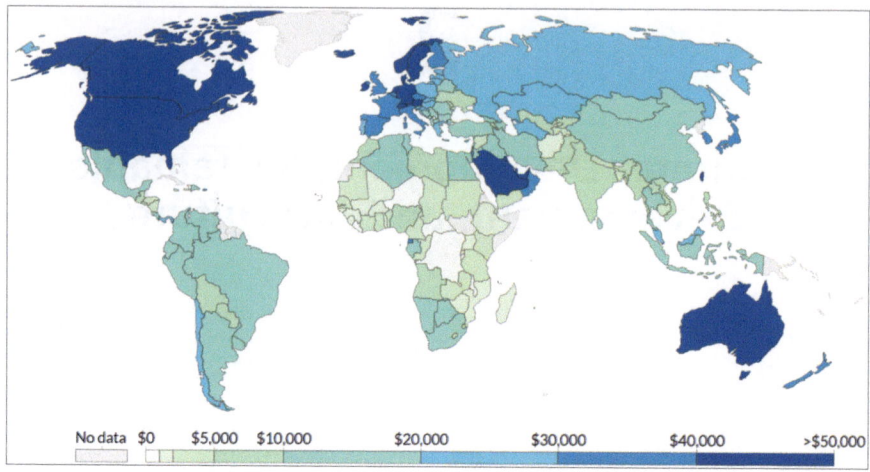

**Fig. 2.1** Gross domestic product (GDP) per capita, 2016. (Source: Maddison Project Database (2018))

countries.' Arguably, the phase of growth and globalization driven by advances in digital technology in the late twentieth and twenty-first centuries marks another major phase of economic growth and associated environmental and social changes (Harari 2017).

The industrial revolution brought about the most rapid period of economic growth and associated improvements in human well-being in history, as measured by 'real gross domestic product per person' (GDP[1] per capita) and a variety of other indicators. This growth produced a set of industrialized countries with high GDP per capita (greater than about US$12,000), a set of transitional or middle-income countries, and a set of low-income or less-developed countries (less than about US$1000), with the largest low-income populations concentrated in sub-Saharan Africa and South Asia (Fig. 2.1).

Figure 2.2 shows GDP per capita for a set of more and less developed countries from the early nineteenth century to early twenty-first century.

---

[1] Throughout this book, we use gross domestic product (GDP) to refer to *real* GDP, that is, GDP measured across countries and over time using prices of a selected base year. For example, the horizontal axis of Fig. 2.1 represents GDP per capita measured using 2011 'international prices' that aim to represent real purchasing power across countries.

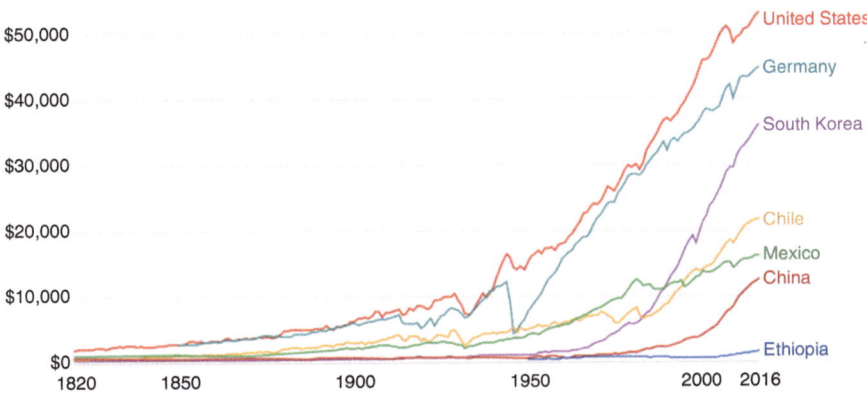

**Fig. 2.2** GDP per capita for selected high- and low-income countries. (Source: Maddison Project Database (2018))

These data show the rapid growth that occurred in what are now the highest income countries, such as the Unites States and Germany, especially in the latter half of the twentieth century. The middle-income countries such as South Korea, Chile, and Mexico also experienced rapid growth but lagged behind the more-industrialized countries and continue to have substantially lower per capita production and incomes. The poorest countries of the world, such as Ethiopia, experienced little growth in per capita GDP in the twentieth century and continue to have incomes that are a small fraction of the high-income countries. China followed a similar pattern until the 1980s when it embarked on a major shift in policy toward a more capitalist economy, after which it experienced one of the highest economic growth rates on record. Yet, today its per capita production and incomes are only about one-fifth of the highest income countries. Figure 2.3 shows per capita GDP by country. The richest countries are concentrated in Europe and North America while the poorest in Africa and South Asia.

Various other socio-economic indicators show similar patterns, for example, Fig. 2.3 shows that life expectancy globally and by regions has been increasing over time in all regions but also is generally higher in countries and regions with higher GDP per capita.

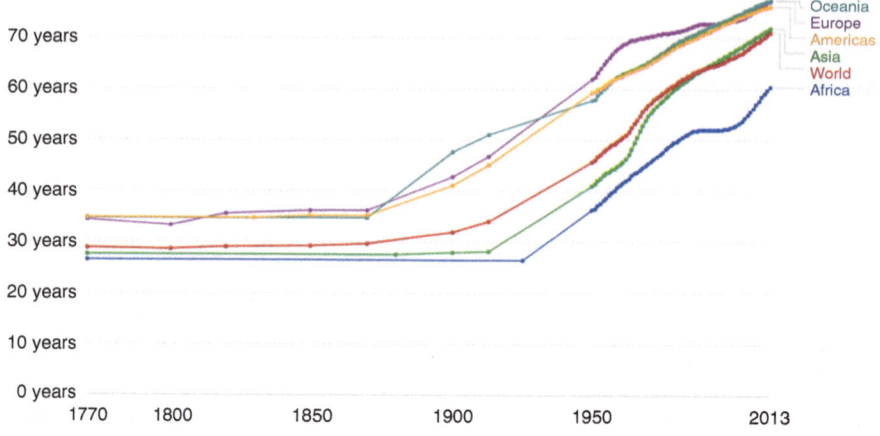

**Fig. 2.3** Life expectancy globally and by world regions. (Source: James C. Riley (2005))

Yet, this remarkable epoch of economic development also brought new social problems associated with the most rapid growth in human population in history, such as the growing obesity epidemic, as well as the most widespread environmental degradation wrought by humans on the natural world. Many of these impacts are local, such as the effects that automobile emissions in urban areas have on air quality and human health. But increasingly, the global expansion of human population and economic growth are having impacts at the global scale, as exemplified by the rapid increase in the concentration of greenhouse gases in the atmosphere and the impact on the global climate (Fig. 2.4). Increasing human dominance of global ecosystems through land transformation and disruption of global water and carbon and nutrient cycles has led some to refer to the twenty-first century as the 'Anthropocene,' the era in which human activity is the dominant influence on climate and the environment (Lewis and Maslin 2015).

The recognition of the environmental challenges created by human dominance of global ecosystems, and the search for solutions, has led to the idea of sustainable development. As we discuss in this chapter, agriculture plays a special role in sustainable development because of its close connection to the land, water, biodiversity, and other natural resources on which all life and human well-being depends.

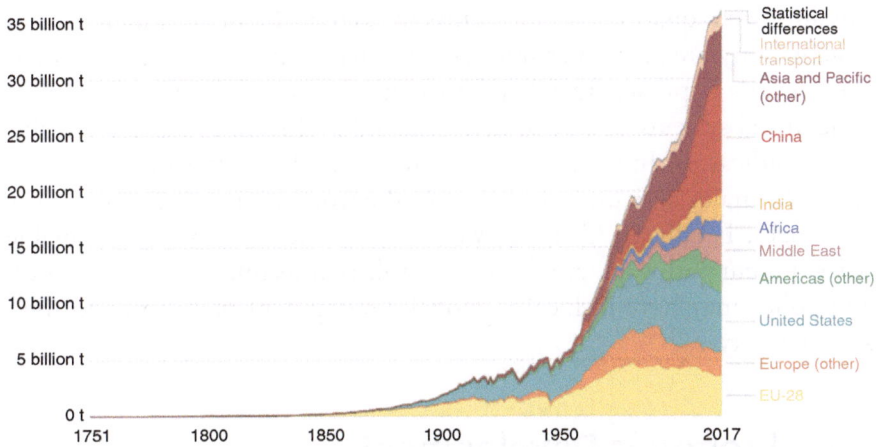

**Fig. 2.4** Annual $CO_2$ emissions (tons/year) by world region. (Source: Global Carbon Budget (2018))

We begin this chapter with a brief discussion of the conventional view of economic growth as the driver of economic and social development and the central role that agriculture has played historically in the growth process, both as a source of labor and other resources and a source of growth essential to the support of expanding human populations. Then, we discuss the more recent extension of the economic and social development paradigm to sustainable development. A key theme of this book is that sustainable development is both a 'positive' conceptual framework for analysis of development and a 'normative' set of goals for development. On the conceptual level, we next describe the three principal dimensions of sustainable development—economic, environmental, and social—and discuss the ways economists have expanded the economic paradigm to accommodate sustainable development and the role that agriculture plays in sustainable development. Then, we turn to the normative aspect of sustainable development and describe the set of sustainable development goals established by the global community. Finally, building on the economic concepts used to analyze sustainable development, we introduce the concept of Tradeoff Analysis as a way to combine the positive and normative approaches to sustainable development. In this approach, quantitative indicators in each of the main dimensions of sustainability are used to characterize alternative development pathways. We use this approach in Chap. 3 to explain how the sustainability of agricultural systems can be understood and evaluated using key sustainability indicators. In Chaps. 4 and 5, we apply the framework to characterize the major agricultural systems of the world and their sustainability challenges. Finally, in Chap. 6, we return to the concept of sustainable development pathways and discuss the role that technological innovation and policy may have in local to global development pathways and their sustainability.

## 2.2   Economic Development

Economic development is a term that came into widespread use in the mid-twentieth century and refers to the process of economic growth and the associated improvements in human well-being as illustrated in Figs. 2.1–2.4. Central to economic development is the process of physical and human capital accumulation that leads to improvements in productivity. Physical capital refers to the various types of plant and equipment used in production, physical infrastructure such as roads and communications, and so on. Although difficult to quantify, economists use the various national accounts statistics collected by countries to estimate the capital stock of each country. For example, according to one set of estimates, the physical capital per capita in the United States in 1980 was in the order of US$100 billion (in 2010 prices) and had increased to about US$160 billion by 2010. In comparison, China's capital stock was estimated to be about US$3 billion per capita in 1980 but increased to about US$50 billion by 2010. Thus, while China had made enormous investment in its capital stock over the thirty years from 1980 to 2010, its capital per person was still only about one-third of the United States (Federal Reserve Bank of St. Louis 2019). As Fig. 2.2 shows, its per capita income in 2010 was approximately one-fifth of the United States.

Another key type of capital investment is 'human capital,' that is, investments in the education, health, and general well-being of individuals that makes them productive members of society. The Human Capital

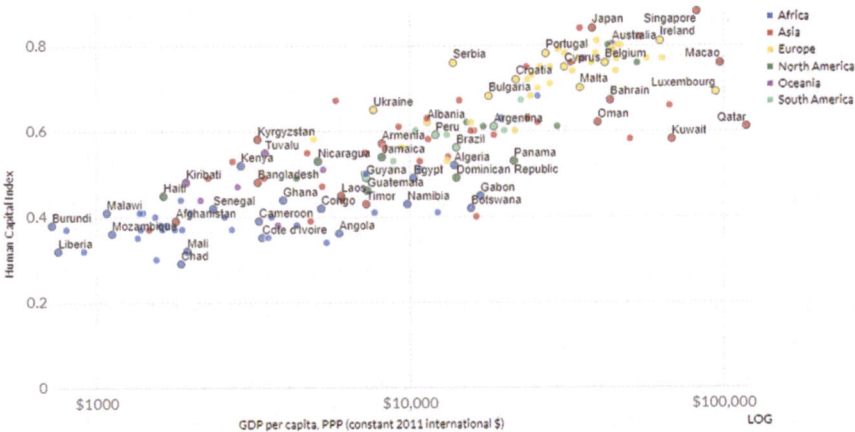

**Fig. 2.5** Human Capital Index versus GDP per capita, 2016. (Source: The World Bank (2019b))

Index combines indicators of child and adult mortality, quantity and quality of education, and health measured in terms of stunting in children and height of adults, to summarize a population's attributes relative to a benchmark of complete education and full health (World Bank 2019a). As Fig. 2.5 shows, the Human Capital Index is positively related to GDP per capita across low and high-income countries. A great deal of research at the individual, regional, and global scales supports the common sense notion that healthier, more educated people are generally more productive and earn higher incomes than people who are less healthy and less well-educated. Most people would also agree that both health and education are important components of well-being and quality of life, so human capital investment benefits people both in terms of their capabilities to be productive and earn income and in the enjoyment of income and consumption.

As we discuss below, another important effect of human capital investment and the related growth in per capita incomes is to increase the value of human time. This increase in turn induces important changes in people's behavior with economic growth and development, for example, it tends to reduce human fertility leading to smaller families and lower population growth. Also very importantly, the increase in the value of people's time changes how people allocate their time in farm households between agricultural production, work in the household for food preparation, child rearing and other 'household production' activities, and work outside the farm and household. These changes in behavior are even more evident among urban households, where relatively high incomes and a high value of time induce people to have fewer but 'higher quality' children with more education and better health care. Higher-income households also spend more of their income on services, such as food preparation, cleaning, and child care, typically performed by family members of lower-income households.

Ultimately, how one evaluates economic, environmental, and social change is a normative or value judgment. For example, the economist Amartya Sen argued for 'development as freedom' by which he meant the enhancement of human capabilities (Sen 1999). Various indicators and measures of economic development are in use by organizations such as the World Bank, now available in its online Open Data (https://data.worldbank.org). GDP per capita is one of the most widely used and readily available indicators for virtually all countries and over long time periods, but it is important to recognize that GDP has substantial limitations as an indicator of human well-being: it does not account for the distribution of wealth or income; it accounts only for market-based transactions and thus does not incorporate non-market effects of human activity on the environment (i.e., impacts on 'natural capital' caused by air and water pollution and on the 'ecosystem services' derived from natural capital), nor does it account for social dimensions of well-being such as health, life expectancy, sense of community, or cultural values. Yet, GDP per capita (and similar aggregate indicators such as national income) is positively related to many dimensions of well-being such as health and education (Fig. 2.5) and thus remains an important summary indicator of economic development despite these limitations. As we will discuss below, it is possible to adjust GDP to incorporate some of these non-market effects on the environment ('green' GDP).

Beginning with the classical economists Adam Smith and David Ricardo, many theories of economic growth and development have been elaborated. In the mid-twentieth century, mathematical models of economic growth were proposed, for example, the model of Solow (1956), which described economic growth as the product of population growth, capital investment, and technological change. A key factor in what Kuznets (1971) described as 'modern economic growth' is technological innovation driven by scientific advances. One of the limitations of the early economic models was their treatment of technological improvements and population growth as 'exogenous' processes that drove economic growth without feedbacks from economic growth to technological change or population growth. Subsequently, economists recognized that human fertility and thus population growth are substantially impacted by real income. More recent growth theories also recognized that innovation is driven by investment in basic and applied science and technology in both the private and public sectors, and this investment in turn depends on a country's economic capacity to support research and development. Research shows that the rate of technological advance is determined by a variety of economic, social, political, and institutional factors that affect investments in science and technology. For example, many scientific advances in the twentieth century were the outgrowth of World War II and the subsequent 'space race' of the Cold War between the Soviet Union and the United States. In Chap. 5, we discuss how technological innovation in agriculture is influenced by economic factors such as resource scarcity. Yet, predicting future trends in research investment, and particularly in predicting individual technological innovations, such as the internet or the iPhone, or the sudden emergence of unmanned aerial vehicles in agriculture, remains largely beyond economics or any scientific discipline.

## 2.3  Role of Agriculture in Economic Growth and Development

Agriculture has long been understood to play a central role in economic growth and development (Mellor 2017; World Bank 2007). Several stylized facts of economic development are important to agriculture and its role in economic growth and sustainable development. One basic fact is that most economic activity is primarily in agriculture before industrialization begins. This was true of eighteenth-century England before the industrial revolution began there and is true of the least-developed countries in today's world. Consequently, the process of industrialization is necessarily a process of moving resources, both physical and human, out of agriculture into other activities.

We can analyze economic growth in terms of ongoing demand-side and supply-side changes occurring in product markets, labor markets, and other resource markets. On the supply side, a key driver of growth is technological innovation, such as the introduction of water-powered textile mills in the industrial revolution in England, which increased supply and lowered real prices of textiles. On the demand side, a key factor is Engel's Law—the idea that as per capita incomes increase along with economic growth, the demand for higher-quality foods, as well as for manufactured goods and for services, increases more rapidly than the demand for basic 'staple' foods and other basic (typically, labor-intensive) goods and services. These changes in demand growth also serve to encourage the production of manufactured goods and provision of services, thus shifting resource use and employment in the economy away from basic agricultural commodity production toward the manufacturing and service sectors.

The shift to industrial production likewise increases the demand for labor and raw materials and raises wages in the industrial sector, thus incentivizing people to migrate from rural areas to urban-industrial areas, while also providing higher incomes to fuel product demand. On the supply side of the labor market, industrial growth increases the demand for skilled labor and thus increases the returns to human capital investment (i.e., education). Improvements in incomes and health lead to increases in labor market participation and life expectancy, and thus population growth increases labor availability and quality.

The net effect of these ongoing changes is to increase the total value of goods and services produced in an economy (GDP) but with growth in the value of manufactured goods increasing faster than the value of agricultural products. The result is another of the key stylized facts of economic growth—the share of GDP produced in the agricultural sector declines relative to manufacturing and services (Fig. 2.6).

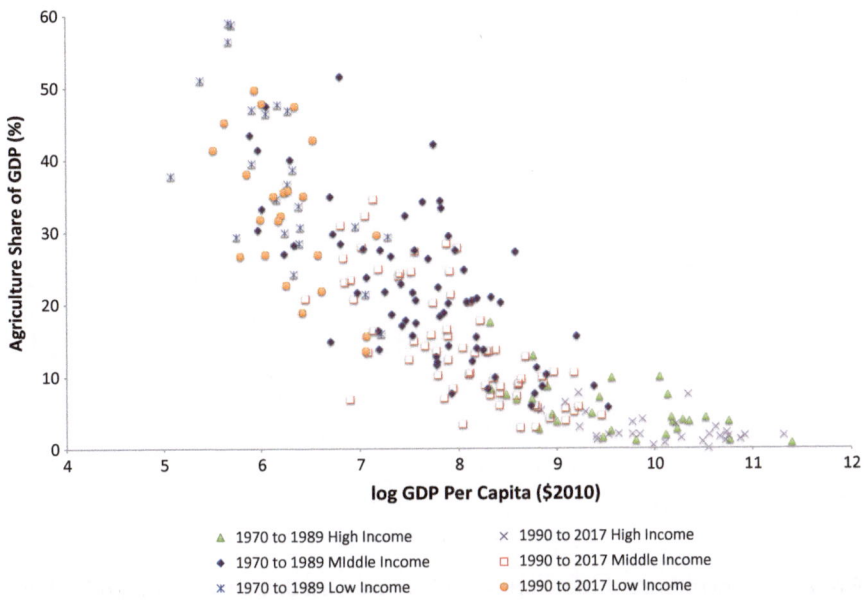

**Fig. 2.6** Agriculture's share of GDP and GDP per capita, 1970–1989 and 1990–2010. (Source: World Bank Data)

As Fig. 2.6 shows, this relative decline in agriculture's importance in growing economies has been true for virtually every country in the world over time as they have developed, and it is true as we compare across more- and less-developed countries at a point in time. However, it is important to note that these changes do not typically lead to an absolute decrease in agricultural production, because productivity growth is also occurring in the agricultural sector during economic growth—what came to be called the 'Green Revolution' of the mid- to late twentieth century. Agricultural productivity growth occurred through the development and adoption of technologies such as mechanization that substitutes for labor and facilitates the movement of labor out of agriculture and also makes possible a large increase in farm size and efficiency. Also fundamental to agricultural productivity growth in the twentieth century was the use of biological technology, such as improved seeds and synthetic fertilizers and pesticides, which increased crop yields. Improved soil and water

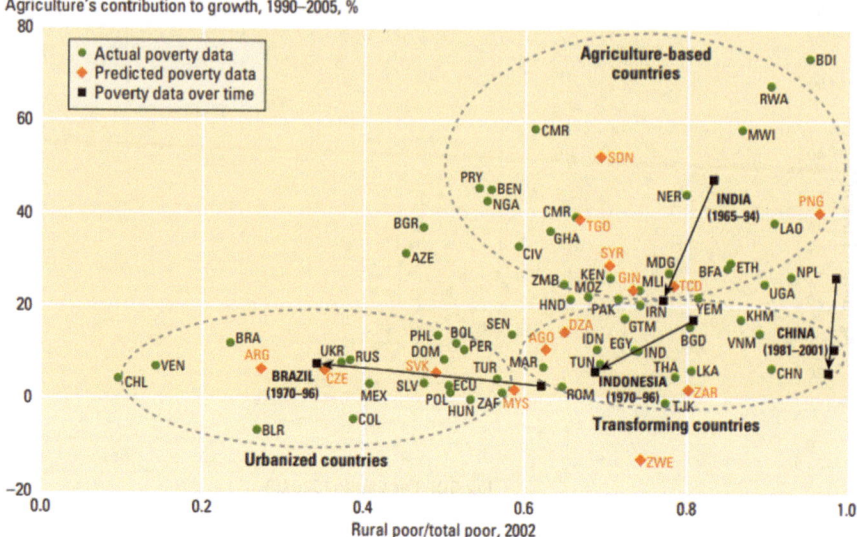

**Fig. 2.7** Agriculture's three worlds: agriculture-based, transforming and urbanized countries. (Source: World Bank (2007))

management and irrigation also contributed to increased productivity. Similarly, improved breeds, feeds, and herd health management improved livestock and dairy productivity.

Another way to summarize the current status of agricultural development globally is presented in Fig. 2.7. Based on analysis by the World Bank, this figure shows that countries can be grouped into three clusters, based on two dimensions: the proportion of the poor households that are located in rural areas and agriculture's contribution to economic growth. Highly agriculture-dependent countries have a large share of the poorest households in rural areas; the transitional economies have a lower dependence on agriculture but still have a high proportion of the poor in rural areas, whereas urbanized (and typically high income) countries have low dependence on agriculture and also a small share of poor households in rural areas.

The late twentieth and early twenty-first centuries saw some other fundamental changes in agricultural and food systems around the world that correspond to these development patterns. Whereas the Green Revolution was focused on the production of enough food calories to feed the rapidly growing world population, in the higher-income regions of the world there is an increasing interest in food quality, nutrition, health, and the environmental dimensions of food production and consumption that we will discuss further in Chaps. 4, 5 and 6. The economic logic that causes agriculture to decline in *relative* importance with economic growth applies to other sectors of an economy as well. Indeed, the higher income countries of the world can be seen to be going through a transition from industrial to postindustrial economies in which the service sector becomes increasingly important. The service sector includes many activities associated with the food system, such as the preparation of food outside the home, as well as financial services, transportation, education and health care, and various other personal services. For example, in the United States, about 30% of personal consumption expenditures were on services in 1960 and that share more than doubled to over 60% by the early 2000s and continues to increase.

In Chap. 6, we discuss further how both demand-side and supply-side factors are playing a role in moving agriculture and the food system toward a development pathway that is more sustainable than the pathway brought about through the technological revolution in agriculture in the twentieth century.

## 2.4    Sustainable Development

As we noted in Chap. 1, we view sustainable development as a conceptual and analytical framework for positive analysis, as well as a set of normative goals. We discuss the normative approach to sustainability in Sect. 2.5. This section describes sustainable development as a conceptual framework founded on the 'three pillars' or dimensions of sustainability: economic, environmental, and social.

## 2.4.1  The Three Dimensions of Sustainable Development

Most analysis of development begins with the assumption that the ultimate goal is improvement of human well-being. Some will argue that this human-centric approach is unjustified, and we leave that debate to the philosophers. Our point is that from the humanist perspective, economic growth plays an obvious fundamental role in meeting and exceeding basic human needs, and this is a widely held value in human societies. This recognition has led to the use of aggregate economic indicators, such as growth in GDP per capita, as barometers of economic development. However, it is now widely recognized that economic performance alone, as measured by aggregated economic performance, is not an adequate metric of human development.

The first reason for a broader view of development derives from the environmental changes that human population growth and economic growth had in the nineteenth and twentieth centuries and the adverse consequences that these changes in turn had on material well-being. These impacts were caused by the conversion of grasslands and forest land for agriculture, extraction of nonrenewable natural resources for materials and energy, and the large-scale transformation of natural ecosystems to managed ecosystems. For example, while expansion of land for agriculture allowed the world to experience unprecedented human population growth in the twentieth century, this growth came at the expense of widespread soil degradation, depletion of groundwater resources in arid regions, contamination of ground and surface water by nutrients and other synthetic chemicals used for pest control, and reductions in beneficial insect populations and biodiversity. These unintended consequences of the agricultural 'Green Revolution' of the twentieth century have had many adverse feedbacks on productivity in agriculture and other industries such as fisheries, as well as direct adverse impacts on human health due to water contamination and exposure to toxic chemicals in the environment and in food.

The second reason for a broader approach to development comes from another widely held value in most human societies that human well-being depends on more than one's own material well-being. Utilitarian philosophy and the conventional economic approach to human behavior suggest one reason for this phenomenon. According to this approach, humans make choices to improve their well-being defined in terms of a metric of value typically called 'utility.' It is often argued that the utility of consuming a good, or more generally of having an extra dollar of income, is increasing but at a decreasing rate—a phenomenon that economists call 'diminishing marginal utility of income.' If this is true, then an additional US$1000 of income for a poor person has much more value or utility than an additional US$1000 of income for a richer person. Indeed, research shows that beyond some level of income at which most material needs are satisfied, people indicate little effect of higher income on their subjective feeling of well-being or 'happiness.'

This principle of a diminishing marginal value of material goods with increasing levels of consumption has important implications for human behavior and for public policy. First, in behavioral terms, it suggests that as real incomes increase with economic growth, humans attach relatively more value to other dimensions of well-being that cannot be bought with money income. As Becker's (1992) research demonstrates, as real incomes rise, the value of people's time increases in economic terms, creating incentives for a wide array of behaviors to save time. Another perspective on this phenomenon is the notion of the 'pursuit of happiness' by enabling individuals' 'capabilities' (Sen 1999), which in turn also provides a rationale for equity in opportunity without regard to gender, race, or ethnicity. In the environmental dimension, many people attach a value to nature for nature's sake—what environmental economists refer to as 'existence value.' This also relates to the value that humans attach to the well-being of other species, including the treatment of wild and domesticated animals consumed by humans. Second, the diminishing value of material goods provides a rationale for policies that redistribute wealth and income in society from rich to poor, both among a society at a point in time and between generations, albeit while taking into consideration the potential distortions that redistributive policies have for incentives to save and invest, both within and between generations.

## 2.4.2 Economic Approaches to Sustainable Development

Beginning in the 1960s, some economists recognized the need to extend conventional economic models of growth to incorporate nonrenewable natural resources such as fossil fuels, renewable resources such as forests and fisheries, and the impacts of human activities on ecosystems and the environment. This work laid the foundation for the more recent extensions of economic models to address sustainable development.

One of the early influential efforts to link economics with environmental concepts was to generalize the conventional 'circular flow' model in economics, which accounts only for market transactions between households and firms, to include natural resource inputs as well as 'residuals' (e.g., greenhouse gas emissions from coal-fired power plants or chemical emissions from manufacturing facilities) into a 'materials balance' analysis (Ayres and Kneese 1969). A related approach was proposed by Weitzman (2007) to generalize the concept of 'national income accounting' on which GDP and related aggregate economic indicators are based. Weitzman's conceptual approach was based on a 'perfectly complete national income accounting' in which both market-based goods and services and 'non-market' goods and services such as the 'ecosystem services' of 'natural capital' are valued and included in the calculation of GDP. This leads to what we could call 'green GDP' or GGDP. In Weitzman's approach, an increase in GGDP implies an improvement in human well-being, and thus a development pathway in which GGDP is increasing would be considered 'more sustainable' than one with GGDP increasing at a lower rate or decreasing.

A key implication of Weitzman's approach is that sustainable development is likely to involve tradeoffs among components of GGDP. For example, early industrial development and urbanization in England involved extensive use of coal for power and heating and led to severe air pollution problems in urban areas such as London but also raised incomes greatly. Further economic development led to alternatives to coal and to cleaner air. If early use of coal had been regulated or banned, it could have meant people would have to forgo the higher incomes that led to subsequent improvements in incomes as well as in environmental quality.

Other researchers tie sustainable development more directly to the idea we introduced above, that economic development is a process of capital accumulation, by generalizing the concept of capital to include natural, human, and intellectual capital as well as conventional physical capital assets such as structures and machinery. Economic sustainability is then defined as maintaining or increasing the total value of all capital stocks (Heal 2016). This approach to sustainability is also related to the approach developed by the World Bank, which attempts to adjust conventional measures of savings and investment to account for changes in all forms of natural, human, and produced capital. As with Weitzman's approach, a key challenge in measuring these changes in capital is to determine appropriate prices or values for non-market components (what economists refer to as 'shadow values').

This capital-theoretic approach to sustainability also leads us to recognize the dynamic aspects of sustainability. One dynamic element is the possibility of irreversible effects and thresholds. For example, if development leads to extinction of some species, we can never regain them in the future. In the context of climate change, warming of the earth system beyond some point may lead to irreversible melting of the Greenland ice sheet and sea level rise.

Conceptualizations of sustainable development by economists such as Weitzman and Heal are useful to help us think beyond conventional economic development. Yet, there are many practical limitations to these abstract economic concepts. As we discuss in Sect. 2.7, the monetary values needed to calculate GGDP or to quantify natural capital are difficult to calculate or justify. Such 'economic' approaches are also politically and ethically challenging because of the difficulty of achieving a consensus on the monetary value of many of the environmental and social dimensions of sustainability. Additionally, even if it could be implemented, an aggregate indicator such as GGDP, or a generalized measure of capital that includes natural capital, obscures the changes and implicit tradeoffs that may be occurring along alternative development pathways. Consequently, we argue in Sect. 2.7 and throughout the remainder of this book for an approach—*tradeoff analysis*—that allows the multiple dimensions of sustainability to be represented explicitly using sustainability indicators measured either in money terms or in other suitable units.

## 2.5 Agriculture's Role in Sustainable Development

Section 2.3 explained why agriculture plays a central role in economic development: first, serving both as a source of land, labor, and other resources that could be transferred to the non-agricultural sectors as they grow; and second, serving as a source of economic growth through technological innovation and productivity growth. As we discussed in Sect. 2.4, in early stages of economic growth and industrialization, most economies follow an unsustainable development pathway, in the sense that we observe environmental quality indicators such as air and water quality declining as GDP per capita increases. Agriculture's role in sustainable development derives from the essential role food plays in human well-being and its close connection to natural resources (land, air, water, biodiversity, energy) and the provision of ecosystem services. In addition to food, agriculture also provides important raw materials such as fiber for clothing, as well as energy in the form of biofuels derived from crops. These interconnections between the economic, environmental, and social dimensions of sustainability are illustrated in Fig. 1.1.

### 2.5.1 A Sustainable Food Supply: Food Security, Nutrition, and Health

Agriculture provides the calories and nutrients necessary for humans not just to survive but to enjoy a healthy and productive life. One expression of these roles is the concept of food security defined at the World Food Summit in 1996 as: 'Food security exists when all people, at all times, have physical, social and economic access to sufficient, safe and nutritious food which meets their dietary needs and food preferences for an active and healthy life.' Elaboration of this definition has led to four generally accepted dimensions of food security: availability, access, utilization, and stability.

- *Food availability*: The availability of sufficient quantities of food of appropriate quality, supplied through domestic production or imports (including food aid).
- *Food access*: Access by individuals to adequate resources (entitlements) for acquiring appropriate foods for a nutritious diet. Entitlements are defined as the set of all commodity bundles over which a person can establish command given the legal, political, economic, and social arrangements of the community in which they live (including traditional rights such as access to common resources).
- *Utilization*: Utilization of food through adequate diet, clean water, sanitation, and health care to reach a state of nutritional well-being where all physiological needs are met. This brings out the importance of nonfood inputs in food security.
- *Stability*: To be food secure, a population, household, or individual must have access to adequate food at all times. They should not risk losing access to food as a consequence of sudden shocks (e.g., an economic or climatic crisis) or cyclical events (e.g., seasonal food insecurity). The concept of stability can therefore refer to both the availability and access dimensions of food security.

When the concept of food security was being developed in the late twentieth century, it was generally associated with the avoidance of undernutrition, first in terms of calories (macronutrients) and then also recognizing the importance of micronutrients. But the increasing prevalence of obesity not only in higher-income countries, but also increasingly in middle- and low-income countries, has shown that we must think of food security more broadly as a problem of *malnutrition* that may also be caused by overconsumption of calories and a lack of dietary diversity. Ultimately, we must recognize that the goal is good health. Understanding the causal linkages from food and nutrition to health shows once again the high degree of complexity and the scientific challenges in achieving a sound basis for understanding sustainable development.

## 2.5.2  Agriculture and the Environment

Agriculture uses more land than any other human activity. Its environmental effects are due, first and foremost, to the fact that land used for agriculture must be transformed in various ways so as to be cultivable: trees and rocks removed; land leveled; roads, irrigation canals, drainage, and other structures built; soil tilled; and its structure, chemistry, and biology modified through additions of organic and inorganic materials and chemicals. Rainfall and irrigation water leach nutrients and chemicals to groundwater, and soil and chemicals run off of farm fields into surface waters. Agricultural activities such as deforestation, land clearing and tillage, flooded rice cultivation, livestock feeding, and use of fossil fuels for power and transport, cause large quantities of the three most important greenhouse gases—carbon dioxide, nitrous oxide, and methane—to be released into the atmosphere. Animal agriculture occupies large areas of land for grazing and feed production, overgrazing can increase soil erosion, and animals emit methane as part of their digestive processes. Animal waste also is a major source of water and air pollution by organic nutrients and bacterial contaminants. Conversion of land to agricultural use is a major cause of biodiversity loss, especially in the tropics. Industrial, mechanized agriculture is also a substantial user of fossil fuels for power and for transport and processing of commodities.

As the reader no doubt is aware, the issues described in the previous paragraph are particularly acute in 'industrial' agriculture and have led to much public debate about their causes, consequences, and policies to address them. Various forms of 'alternative' agriculture have been developed and advocated to address the environmental problems associated with industrial agriculture, including conservation agriculture, organic agriculture, ecological agriculture, and sustainable agriculture. As we noted in Chap. 1, we distinguish these concepts from sustainable agricultural *development* by which we mean a *process* of moving agricultural development along a pathway toward options that recognize and balance tradeoffs among the economic, environmental, and social dimensions of sustainability. We discuss the environmental challenges facing industrial agriculture in Chap. 5.

### 2.5.3 Agriculture and Climate Change

Climate change merits special attention in the discussion of agriculture and the environment because climate change is a major existential threat to humankind, and because agriculture is a major source of greenhouse gas emissions—estimated to be in the range of 10–15% of total emissions globally, with the largest emitting countries China, India, Brazil, and the United States, and the largest source coming from methane emissions of livestock production, followed by nitrous oxide emissions from fertilizer use, and carbon dioxide emissions from land use change and fossil fuel use (Food and Agriculture Organization 2019). The reader can find a huge literature online about climate change and agriculture, including the periodic 'assessment reports' by the Intergovernmental Panel on Climate Change, that summarize the scientific literature and its conclusions. Agriculture plays an important role as an emitter of the three main greenhouse gases (carbon dioxide, nitrous oxide, and methane), and it can also play a role in reducing greenhouse gas concentrations through the process known as carbon sequestration. Carbon can be removed from the atmosphere through photosynthesis and stored in soil and in biomass of plants. We discuss both of these aspects of climate change in later chapters.

### 2.5.4 Agriculture and Gender Equity

In addition to the food security, nutrition, and health aspects of agriculture and the food systems, agriculture has other social dimensions. Agriculture employs a large share of labor in the poorest countries and influences families' decisions regarding education as well as health. Many aspects of gender equity are related to the way that agricultural systems are managed: men and women often perform different work tasks; men and women are involved (or not) in management decisions in different ways; and men and women benefit (or not) from agricultural activities, income, and the control over household assets.

Despite the popularization of Becker's (1992) concept of household production as an important set of economic activities, women's work in the agricultural sector of developing countries is grossly undervalued and underestimated. In many societies, women are primary care givers in a household for children, elderly, and the sick. Particularly in developing countries, women and girls are engaged in necessary household work such as water and fuel collection, processing and preparation of food, livestock management, and several other household chores. These activities constrain women's participation in paid work. When women are hired as agricultural laborers, there is a substantial wage gap, and their bargaining power is limited by cultural norms. These norms also limit mobility of women in many societies, so men are more likely to migrate to urban centers for higher-paying jobs in the non-agricultural sector. Male out-migration from agriculture increases women's responsibilities for on-farm work and coping with adverse economic or weather shocks to the household. The phenomenon has been described as 'feminization' of agriculture. However, due to women's limited control over productive assets and inferior social status, leading to discrimination in access to irrigation, credit, and extension services, female-headed households are often less productive. With the advent of improved communication technologies using mobile phones, migrant male family members are also able to exert greater financial control over the proceeds of women's work.

A growing body of literature has recognized that male and female household members may have divergent preferences over the use of household resources (Agarwal 1985; Quisumbing and Maluccio 2000). Women are more likely to allocate resources under their control to food, education, and health care (Meinzen-Dick et al. 2012; Cunningham et al. 2015; Brunson et al. 2009). Given the crucial role of women in every aspect of food and nutritional security including production, preparation, allocation, consumption, and child care, increased autonomy of women over households resources can have far-reaching impact on the pursuit and achievement of several Sustainable Development Goals (SDGs) (No Poverty, Zero Hunger, Good Health, and Well-Being to mention a few) discussed in the following section.

Box 2.1  **Women's Control over Agricultural Land in India**

Bina Agarwal's research and advocacy on gender and land rights has high-lighted that control over agricultural land is most important to improve women's bargaining power in rural economies. She traces the slow progress in increasing women's control over agricultural land in India (Agarwal 2003). Land transfers during land reforms in India were made to male members of a household, which counted an adult son as a separate unit but not an unmarried adult daughter. Even in female-headed households, land was often allocated to the woman's son instead of the female household head. Agarwal (2003) lists four reasons that underlie widespread biases in favor of male control over land—(1) men are usually perceived as earning members in a household and women are dependents; (2) women's role in patrilineal societies limits their ability to directly cultivate land inherited from their families due to rigid gendered social norms; (3) a household is often incorrectly considered a single unit that shares common interests and preferences; and (4) a deep-seated unfortunate belief that women's control over property can destabilize harmony within a household.

The true extent of inequality in land ownership is difficult to estimate given the lack of large-scale survey efforts to record such information. However, long-standing social and administrative biases imply that actual control over property is quite different from legal rights and ownership on paper. In many cases, women forego their share of land inheritance in favor of their brothers to maintain an effective source of security by conforming to social norms. There are some regional variations in these social norms. While southern India has seen some relaxation, regions in north and north-western India continue to impose rigid social restrictions on women's rights.

## 2.6 Pathways to Sustainable Development: The Sustainable Development Goals

In Chaps. 4 and 5, we discuss the characteristics and sustainability challenges of developing, transitional, and industrial agricultures, as well as some options for moving the current systems in more sustainable directions. We argue in Chap. 6 that while individual actions can help achieve sustainable development goals, substantial changes in development pathways will require a deliberative process to identify goals and move underlying trends in technology, economic and social drivers, and processes toward more sustainable pathways. Sachs (2015) usefully compares what he calls a 'business-as-usual' (BAU) pathway—the pathway the world is now on, and recognized as unsustainable—from a 'sustainable development' (SD) pathway. One key feature of the BAU pathway is that it is almost entirely dependent on fossil fuels as a source of energy, despite ongoing efforts to develop alternative, more sustainable energy sources. It is clear that continuing on this development pathway will result in large and unsustainable increases in global temperatures and other environmental changes that will threaten humanity in many dimensions (Intergovernmental Panel on Climate Change 2018). Thus, it is clear that one feature of a SD pathway will involve a transition toward forms of energy that will allow the concentration of greenhouse gases in the atmosphere to be stabilized at levels near those of today. As we discuss in Chap. 6, there are good reasons to think that some combination of technological, social, and political changes will be required, supported by public policies at local, national, and global levels, to achieve this energy transition. This will likely be true for achievement of the various other sustainable development goals that have been proposed by the United Nations, including the transition of agriculture toward a more sustainable development pathway.

Every individual and every organization may have their own ideas and values about what constitutes sustainable development goals—that is what we mean when we say the SD goals are *normative*. However, in order for individuals and organizations from local to global levels to work cooperatively and effectively, some agreement on a set of common goals is needed. For this reason, a set of sustainable development goals (SDGs) were discussed and adopted at the UN Sustainability Development Summit in 2015 (Fig. 2.8). The SDGs refer to 17 core targets of the 2030 Agenda for Sustainable Development which were developed, building on the millennium development goals (MDGs) set in 2000 for the year 2015. At the 2012 Rio Earth Summit, world leaders recognized there had been mixed progress toward the MDGs. While there was substantial success in some economic and social dimensions—notably, in reducing extreme poverty in some parts of the world such as China, and in increasing access to primary school and reducing the spread of preventable diseases during the first decade of the twenty-first century—it was clear that the path of economic development was far from sustainable in many other social and environmental dimensions.

Fig. 2.8 Sustainable development goals (SDGs). (Source: United Nations, https://www.un.org/sustainabledevelopment/. Note: The content of this publication has not been approved by the United Nations and does not reflect the views of the United Nations or its officials or member states)

As we noted above, the challenge of climate change is one of the most obvious examples of the BAU pathway the world is on. Since the Kyoto Protocol was adopted in 1995, reports by the Intergovernmental Panel on Climate Change have highlighted the increases in greenhouse gas emissions, warming of the atmosphere and oceans, loss of biodiversity, and their likely current and future impact on the three dimensions of sustainability (Intergovernmental Panel on Climate Change 2014). Recognizing the success of MDGs to motivate global action and the interlinked challenges in the pursuit of sustainable development, seventeen SDGs were agreed upon as 'a short set of compelling objectives' based on 'The Future We Want' vision of global leaders. While the goals are broad, they are further elaborated using indicators and targets for each of the goals. These indicators enable continuous tracking and evaluation of progress toward these goals, which can be disaggregated to regional and national levels.

The SDGs and the associated indicators and targets are guided by the three pillars of development—economic, environmental, and social—

and are thus inherently multidimensional. Each Goal is quantified with targets based on multiple indicators documented in the SDG Handbook (https://www.un.org/development/desa/capacity-development/tools/tool/e-handbook-on-sustainable-development-goals-indicators/). For example, the first goal of 'No Poverty' aims to eradicate extreme poverty and improve resilience of the vulnerable population. Achievement of the Goal 1 is based specifically on five Indicators: percentage of population living under in extreme poverty (i.e., less than US$1.90/day); coverage of social safety nets; the impact of disasters in terms of the preparedness of countries; the number of people affected; and economic loss as a result of disasters. The SDG Handbook provides a detailed discussion of each goal and the associated targets. Many of them highlight tradeoffs within the agricultural sector or between agriculture and other sectors in order to meet the overarching goals.

The first three SDGs ('No Poverty,' 'Zero Hunger' and 'Good Health and Well-Being') together aim to improve human development and are directly linked to the agricultural sector, which employs 25% of the global population. But as we noted earlier in this chapter, an even larger share of the population depends on agriculture and allied sectors for their livelihoods. The FAO estimates that, by 2050, food production will need to increase by 70% to keep up with the increase in population and demand for food, largely from developing regions. Eradication of extreme poverty will require rapid increases in the productivity of the agricultural sector. Goal 12 (Responsible Production and Consumption) and Goal 13 (Climate Action) will require that the first three goals are achieved with minimum environmental impact in the present and preserving the productive capacity of natural resources for future generations. In Chaps. 3, 4 and 5, we discuss indicators for agricultural sustainability in greater detail and their correspondence to the SDGs.

## 2.7    Tradeoff Analysis of Sustainable Development Pathways

In this section, we describe the use of 'tradeoff curves' to analyze and visualize sustainable development options using sustainable development indicators, such as those associated with the SDGs and the ones elaborated for agriculture in Chap. 3. To make analysis and visualization tractable, we will typically select a small number of key indicators in the economic, environmental, and social dimensions.

## 2.7.1 Defining Tradeoff Curves

Tradeoff curves derive from some basic concepts and principles of economics. In economic theory, a 'transformation frontier' or 'production possibilities frontier' is used to represent the tradeoffs that exist in an economy as productive resources (land, labor, capital, energy, materials) are allocated among alternative productive uses. In the simplest case, the activities of an economy are represented by two sectors, such as agriculture and manufacturing, whose outputs can be represented on a graph such as Fig. 2.9. For example, Indicator 1 could measure manufacturing output, and Indicator 2 could represent agricultural output in an economy. This frontier shows the possible combinations of the two sectors' outputs that the economy can produce as resources such as labor and capital are shifted from one sector to another. For example, if labor were to migrate out of rural areas to cities, and thus move from being employed in agriculture to manufacturing, agricultural output would decrease and

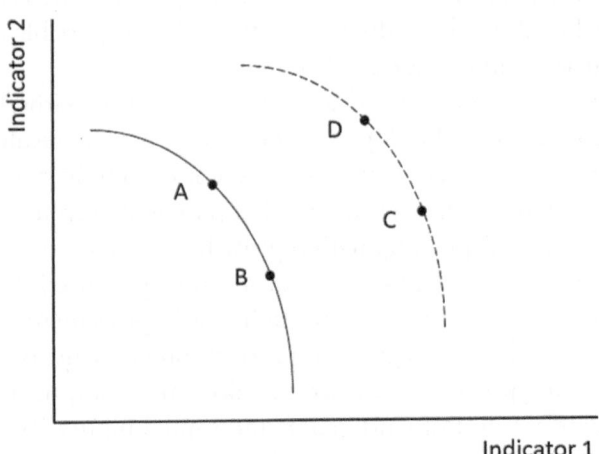

**Fig. 2.9** Transformation frontiers and tradeoff curves

manufacturing output would increase as indicated by a move such as from points A to B in Fig. 2.9. Note that we use two sectors so we can represent it graphically. More generally, we can have many sectors and can define the transformation frontier among all of them mathematically, but we cannot represent it graphically.

The movement of resources from one sector to another usually cannot be made on a one-for-one basis, however, at least beyond some point. For example, labor trained to work in agriculture may be less productive in manufacturing. Additionally, as increasing numbers of workers are moved from agriculture to manufacturing, their 'marginal productivity' in manufacturing declines relative to agriculture (all else equal). Thus, economic theory shows that the transformation frontier, which represents the maximum output obtainable in each sector from different combinations of available resources allocated to the two sectors, has the shape shown in Fig. 2.9, concave to the origin.

The transformation frontier illustrated in Fig. 2.9 is constructed for a given technology (way of producing) in each sector and a given total resource endowment in the economy. Given these factors—technology and resource endowment—the frontier shows the various possible combinations of agricultural and manufacturing output that are obtainable from *efficient* use of the available resources. Importantly, if some resources are not used (e.g., if some labor is not employed), the economy represented in Fig. 2.9 will produce less output than is possible and so will be operating at a point inside the frontier.

Another key economic idea is that changes in either the resource endowment or the technologies in the two sectors will result in a different transformation frontier. Thus, we represent economic growth as an outward 'shift' in the frontier, showing that potential output in the economy increases. For example, population growth increases the amount of labor in the economy and allows more output to be produced. But an increase in productivity, for example, through the improvement in the genetic potential of seeds used to grow crops, or through the use of more efficient manufacturing processes, also would allow more output to be produced for the same amount of land, labor, and capital inputs. This type of shift in the transformation frontier is illustrated in Fig. 2.9, by the dashed frontier. An economy initially at point A can then move to a new point such as C or D, where more agricultural output or more manufacturing output, or more of both, can be produced with the same resource endowment of land, labor, capital, energy, and materials.

The transformation frontier demonstrates two fundamental concepts in economics, which we will use in our analysis of sustainable development. First, for *given* resources and technologies, and assuming efficient use of those resources, in order to obtain more of one output, we must give up some amount of another output—this is represented by the negative slope of the frontier. The key implication is that there are inevitable tradeoffs between alternative desired outcomes in the economy. In economics, this principle is codified in the concept of *opportunity cost*, that is, to get more of one thing, something else must be given up (economists often say, 'there is no free lunch'). Second, an economy can produce more output in three ways: (1) if production is inefficient (there are unemployed resources, or production methods are not being utilized appropriately), reducing this inefficiency can move the economy toward the frontier, but if we are using resources and technologies efficiently and thus are on the frontier, then the only way to increase output in multiple sectors is to shift the frontier outward by (2) increasing the amount of resources available or (3) improving the productivity potential of the production technologies.

Using these concepts, we can describe economic development as the process of moving an economy to different points in Fig. 2.9. Economic growth typically results in the movement along a pathway such as from points A to D or B to C, providing an increase in both sectors of the economy. The particular pathway depends on both the supply side of the economy, represented in Fig. 2.9 by the transformation frontier, and the demand side of the economy and the interaction of supply and demand through markets, as well as government activities that also affect resource allocation through taxes, subsidies and government spending.

### 2.7.2  Using Tradeoff Curves to Analyze Sustainable Development Pathways

Now we can generalize the concept of a production possibilities frontier to describe development pathways in relation to sustainability. We will now use Fig. 2.9 to represent *tradeoff curves*, which show the possible combinations of economic, environmental, and social outcomes, as represented by *sustainability indicators*, such as those associated with the SDGs. Note that we define these tradeoff curves for a particular state of the world, analogous to the technologies and resource endowments that define the transformation frontier discussed in the previous section. By state of the world, we mean the resources, technologies, policies, and other factors that determine the feasible combinations of indicators. We expect in most cases that, given the state of the world in terms of technologies for producing food and other marketed products, an increase in economic activity will tend to result in some environmental degradation, so we have a negatively sloping tradeoff curve between economic and environmental indicators. In social dimensions, it is less obvious that there is necessarily a tradeoff; for example, increased economic activity could be associated with better health or a more equal income distribution. Thus, in some cases, win-win outcomes are possible and the curve will have a positive slope. Yet, there are also surely many outcomes where there are tradeoffs in the social dimension as well; for example, where economic growth is associated with poorer health due to increased stress or exposure to pollution caused by industrial activities.

To illustrate, let Indicator 1 represent an economic indicator such as GDP per capita and let Indicator 2 represent an environmental indicator such as air quality, or a social indicator such as equality of income distribution. Thus, a change in the economy is likely to result in a movement along the tradeoff curve (such as from points A to B). In the environmental dimension, for example, there is ample evidence that developing countries tend to experience environmental degradation with economic growth. Yet, as development progresses, there is also evidence of environmental improvements, for example, as people with higher incomes seek a higher quality of life and support policies for environmental protection and development of cleaner technologies. The implementation of environmentally friendly policies and technologies represents a change in the state of the world that makes it possible for the economy to shift the tradeoff curve outward and move from a point like B to a point like D in Fig. 2.9. This pattern of a movement such as from points A to B to D has been seen in many countries. A curve tracing out this pattern is sometimes referred to as an 'environmental Kuznets curve' in honor of the relationship between equality of income distribution and economic development first observed by economist Simon Kuznets.

As with conventional economic development, there are many possible development pathways that could be more or less sustainable—by which we mean following a positive or negative trajectory in Fig. 2.9. For example, the movement from A to B to C results in a net loss in the environmental or social dimension (the vertical axis) as well as an improvement in the economic dimensions (horizontal axis). However, a more ambitious effort to improve policy or technology could also lead to a movement with improvements in both dimensions, such as from point A to D, thus representing a more sustainable development pathway.

Of course, actual development pathways are much more complex with many dimensions (as with the SDGs), too many to visualize graphically in two or even three dimensions.

## 2.8    Conclusions

In this chapter, we have introduced the key role that agriculture plays in economic growth, economic development, and in sustainable development. We also introduced the sustainable development goals as a way to represent the normative side of sustainable development, and tradeoff analysis as a way to carry out the positive analysis of sustainable development, using indicators for the economic, environmental, and social dimensions of sustainability. In Chap. 3, we use these concepts to focus on the sustainability of agricultural systems as the foundation of sustainable agricultural development.

## References

Agarwal, Bina. 1985. Work Participation of Rural Women in Third World: Some Data and Conceptual Biases. *Economic and Political Weekly* 20: A155–A164.

———. 2003. Gender and Land Rights Revisited: Exploring New Prospects Via the State, Family and Market. *Journal of Agrarian Change* 3 (1–2): 184–224.

Ayres, Robert U., and Allen V. Kneese. 1969. Production, Consumption, and Externalities. *The American Economic Review* 59 (3): 282–297.

Becker, Gary S. 1992. Nobel Price Lecture. Nobel Media AB 2019. https://www.nobelprize.org/prizes/economic-sciences/1992/becker/lecture/.

Brunson, Emily K., Bettina Shell-Duncan, and Matthew Steele. 2009. Women's Autonomy and Its Relationship to Children's Nutrition Among the Rendille of Northern Kenya. *American Journal of Human Biology: The Official Journal of the Human Biology Council* 21 (1): 55–64. https://doi.org/10.1002/ajhb.20815.

Cunningham, Kenda, George B. Ploubidis, Purnima Menon, Marie Ruel, Suneetha Kadiyala, Ricardo Uauy, and Elaine Ferguson. 2015. Women's Empowerment in Agriculture and Child Nutritional Status in Rural Nepal. *Public Health Nutrition* 18 (17): 3134–3145. https://doi.org/10.1017/S1368980015000683.

Federal Reserve Bank of St. Louis. 2019. University of Groningen and University of California, Davis, Capital Stock at Constant National Prices for China [RKNANPCNA666NRUG]. Accessed May 12, 2019, from FRED. https://fred.stlouisfed.org/series/RKNANPCNA666NRUG.

Food and Agriculture Organization. 2019. www.fao.org/faostat/en/#data/GT/visualize.

Global Carbon Budget. 2018. *Earth System Science Data* 10 (4): 2141–2194.

Harari, Y.N. 2017. *Homo Deus: A Brief History of Tomorrow*. Harper-Collins.

Heal, Geoffrey. 2016. *Endangered Economies—How the Neglect of Nature Threatens Our Prosperity*. Columbia University Press. https://cup.columbia.edu/book/endangered-economies/9780231180849.

Intergovernmental Panel on Climate Change. 2014. *Climate Change 2014: Impacts, Adaptation, and Vulnerability. Part A: Global and Sectoral Aspects*. Contribution of Working Group II to the Fifth Assessment Report of the Intergovernmental Panel on Climate Change. Cambridge University Press. https://www.ipcc.ch/site/assets/uploads/2018/02/WGIIAR5-PartA_FINAL.pdf.

———. 2018. *Summary for Policymakers*. In Global Warming of 1.5°C. An IPCC Special Report on the Impacts of Global Warming of 1.5°C above Pre-industrial Levels and Related Global Greenhouse Gas Emission Pathways, in the Context of Strengthening the Global Response to the Threat of Climate Change, Sustainable Development, and Efforts to Eradicate Poverty [Masson-Delmotte, V., P. Zhai, H.-O. Pörtner, D. Roberts, J. Skea, P.R. Shukla,

A. Pirani, W. Moufouma-Okia, C. Péan, R. Pidcock, S. Connors, J.B.R. Matthews, Y. Chen, X. Zhou, M.I. Gomis, E. Lonnoy, T. Maycock, M. Tignor, and T. Waterfield (eds.)]. World Meteorological Organization, Geneva, Switzerland, 32 pp.

Kuznets, Simon. 1971. Nobel Prize Lecture. Nobel Media AB 2019. https://www.nobelprize.org/prizes/economic-sciences/1971/kuznets/facts/.

Lewis, Simon L., and Mark A. Maslin. 2015. Defining the Anthropocene. *Nature* 519 (7542): 171–180. https://doi.org/10.1038/nature14258.

Maddison Project Database, Jutta Bolt, Robert Inklaar, Herman de Jong, and Jan Luiten van Zanden. 2018. Rebasing 'Maddison': New Income Comparisons and the Shape of Long-run Economic Development. Maddison Project Working paper 10.

Meinzen-Dick, Ruth, Julia Behrman, Purnima Menon, and Agnes Quisumbing. 2012. Gender: A Key Dimension Linking Agricultural Programs to Improved Nutrition and Health. In *Reshaping Agriculture for Nutrition and Health*, 135–144. Washington, DC: International Food Policy Research Institute.

Mellor, J.W. 2017. *Agricultural Development and Economic Transformation: Promoting Growth with Poverty Reduction*. Palgrave Macmillan.

Quisumbing, Agnes, and John A. Maluccio. 2000. Intrahousehold Allocation and Gender Relations: New Empirical Evidence from Four Developing Countries. Washington, DC: IFPRI.

Riley, James C. 2005. Estimates of Regional and Global Life Expectancy, 1800–2001. Issue Population and Development Review. *Population and Development Review* 31 (3): 537–543. [Zijdeman, Richard, and Filipa Ribeira da Silva. 2015. Life Expectancy at Birth (Total). http://hdl.handle.net/10622/LKYT53. IISH Dataverse, V1, and UN Population Division (2019).]

Sachs, Jeffrey D. 2015. *The Age of Sustainable Development*. New York and Chichester, West Sussex: Columbia University Press.

Sen, Amartya. 1999. *Development as Freedom*. New York: Oxford University Press.

Solow, Robert M. 1956. A Contribution to the Theory of Economic Growth. *The Quarterly Journal of Economics* 70 (1): 65–94. https://doi.org/10.2307/1884513.

Weitzman, Martin L. 2007. *Income, Wealth, and the Maximum Principle*. Cambridge, MA and London: Harvard University Press.

World Bank. 2007. World Development Report 2008: Agriculture for Development. Washington, DC: World Bank. https://openknowledge.worldbank.org/handle/10986/5990

———. 2019a. World Development Report 2019: The Changing Nature of Work. Washington, DC. Accessed November 17, 2019. http://documents.worldbank. org/curated/en/816281518818814423/pdf/2019-WDR-Report.pdf.

———. 2019b. Human Capital Index vs. GDP per Capita. Our World in Data. https://ourworldindata.org/grapher/human-capital-index-vs-gdp.

World Food Summit. 1996. Rome Declaration on Food Security and World Food Summit Plan of Action. FAO. http://www.fao.org/docrep/003/ w3613e/w3613e00.HTM.

# 3

# Sustainability of Agricultural Systems

## 3.1 Introduction

In Chap. 2, we discussed agriculture's role in economic development and sustainable development and the aggregate indicators that can be used to set sustainable development goals and evaluate progress toward them. We concluded by describing tradeoff analysis as a way to analyze sustainable agricultural development by evaluating alternative development pathways into the future.

In this chapter we bring the concept of sustainability to the analysis of agricultural systems. We describe agriculture as a diverse array of production systems that are composed of interconnected physical, biological, and human components. Our key insight is that to improve the sustainability of agricultural systems, we must have a scientific understanding of how their components work and interact over space and time. As we discussed in Chap. 2, in the aggregate context of sustainable development goals, although it is desirable to improve performance in all three dimensions of sustainability—economic, environmental, and social—this is often not possible, so we must contend with tradeoffs. We shall see this more concretely at the farming systems level in Chaps. 4 and 5 when we discuss the

© The Author(s) 2020
J. M. Antle, S. Ray, *Sustainable Agricultural Development*, Palgrave Studies in Agricultural Economics and Food Policy, https://doi.org/10.1007/978-3-030-34599-0_3

complexity of the many agricultural systems around the world and their interactions with wider natural and human systems. There are almost always tradeoffs among the three dimensions of sustainability in individual agricultural systems, and it is these system-specific features that give rise to the aggregate tradeoffs we discussed in Chap. 2. As we elaborate in Chap. 6, our strategy for improving the sustainability of agricultural development pathways is to quantify and understand tradeoffs and synergies among the dimensions of sustainability, and how both demand-side and supply-side factors in an economy can influence those pathways.

Two key features of agricultural systems that we highlight in this and following chapters are their *diversity* and their *heterogeneity*. By diversity, we mean the different types of systems that exist across the regions of the world. Beginning in this chapter, and continuing through Chaps. 4, 5 and 6, we use two examples that illustrate diversity and heterogeneity: a semi-subsistence farm household system typical of many regions of the developing world and a commercial farm household system typical of the industrialized world. In this chapter, we discuss characteristics of agricultural systems and then use them to describe and evaluate major systems in developing regions (Chap. 4) and industrialized regions (Chap. 5).

We use the term 'heterogeneity' to describe the variation in the physical, biological, and human components *within* each type of the system and the resulting variation in their sustainability properties. Data from around the world show that there is a tremendous amount of heterogeneity in agricultural systems, and science has demonstrated that heterogeneity plays a key role in understanding the processes that determine a system's performance. Some of these processes operate on very small scales, for example, crop growth and its relationship with soil fertility and the farmer's crop management may operate at the individual plant or field scale on a daily basis. However, other processes operate at much larger spatial and temporal scales, for example, the movement of nutrients off farm fields and their transport to larger water bodies may involve large areas of land and many months or even years. Likewise, policy decisions are typically influenced by economic, environmental, or social outcomes not at an individual or household level but for larger regions or groups. Thus, our approach to agricultural system sustainability necessarily involves an understanding of systems at the farm level as well as larger scales that we refer to as landscapes and populations.

Taken together, our goal in this and following chapters is to explain how the diversity of agricultural systems and their heterogeneity over space and time give rise to the challenges humanity faces in achieving more sustainable agricultural development pathways as an essential part of achieving more sustainable development in economies and societies.

## 3.2 Agriculture as Complex Managed Systems

Science teaches us that the world is composed of many interconnected *complex systems*, as illustrated in Fig. 1.1. Various terminologies are used to describe these systems, but one important distinction is that agriculture is a part of *managed* systems, in contrast with what is often referred to as 'natural' or unmanaged systems. With the growth and geographic spread of human populations, there are now few, if any, truly natural, unmanaged ecosystems in the world. For example, ecosystems in protected areas such as national parks are in a sense managed to produce a set of 'ecosystem services,' such as habitat to support wildlife and biodiversity conservation, achieved by minimizing the impact of humans on those ecosystems. In contrast, the principal goal of agricultural systems is to produce food, fiber, and fuel products for human consumption.

This way of looking at agriculture helps us understand both the positive and normative challenges of sustainable agricultural development that we shall discuss further in the rest of this book. The degree to which human impact on 'natural' as well as 'managed' ecosystems is desired is a question of degree and the values of the humans managing the systems. For example, in the United States, many people think that some mountainous regions should be maintained in a natural, undisturbed condition as much as possible by restricting most human uses of the land, such as livestock grazing, mechanized travel, or permanent human habitation. In Switzerland, in contrast, many people think that traditional mountain pastoral areas created by people living there centuries ago for livestock production should be managed so as to prevent them from reverting to their original forested, 'unmanaged' condition. Neither of these approaches is right or wrong but rather reflects the normative feature of sustainable development goals and underlines the fact that different people in different places and cultures may have different values and goals. But in both cases, achieving the desired goals through purposeful management requires an understanding of the complex systems that are involved.

As we noted in the introduction, agricultural farm household systems operate at a wide range of spatial and temporal scales, interacting with other bio-physical, economic, and social systems at various scales. For example, living organisms such as a corn plant are made up of cells with a particular genetic makeup (genotype) that usually has been modified by humans, either through conventional plant breeding or through modern laboratory techniques (biotechnology). A field of corn cultivated by farmers is made up of many corn plants, and an agricultural region such as the US Midwest contains many fields of corn. Likewise, these corn plants are rooted in the soil and interact with the hydrologic system that itself is composed of a set of interconnected surface and groundwater components. Corn plants also interact with the atmosphere through photosynthesis and evapotranspiration. When a farmer applies nutrients to the soil in a field of corn, some of those nutrients are taken up by the plants, some are transformed into a gas (nitrous oxide) and enter the atmosphere as part of the global nitrogen cycle, and some nutrients runoff into surface water and are carried through the interconnected network of streams and rivers to the US Great Lakes and the Gulf of Mexico. The large amount of nutrients transported from farm fields to the Gulf and Great Lakes contributes to oxygen depletion (called hypoxia) and in turn impacts aquatic life in the Gulf and the Lakes. Corn production provides livelihoods to farm households and is a major component of the regional economy in the US Midwest and ultimately to the national economy. Much of the corn production in the US Midwest is exported to other countries to be fed to livestock, thus affecting global prices, meat production and consumption, and the associated health and environmental outcomes in other parts of the world. Some of those nutrients taken up by the corn plant are stored in the corn grain, are eaten and digested by livestock in China, and become animal waste that pollutes air and water in China. The meat produced also contributes to the nutrition of Chinese people.

As this example illustrates, farms themselves are complex systems composed of many components: physical (soil, climate), biological (crops, livestock, soil microorganisms, insects, birds, etc.), and human (farm household, farm workers, rural residents). Figure 3.1 provides an illustration of a farm household system portrayed as a household production component on the right-hand side and a farm production component on the left-hand side. From an economic perspective, we can think of the farm and household as interconnected production systems with inputs and outputs. Increasingly in the industrialized regions of the world, the commercial farms that produce a large share of agricultural commodities are businesses, often organized as partnerships or corporations. Like all industrial firms, these farms invest in capital (land, machinery, and structures), employ labor, purchase inputs (e.g., seed, fertilizer, fuel), and produce outputs (e.g., crops and livestock). While these commercial farms have primarily economic motivations and objectives, it is likewise true that 'family' farms must be viable economically to continue operating in the long term. The exception to this rule are the 'hobby farms' operated by people who have other sources of income and choose to operate a farm that loses money but provides the owners with other 'lifestyle' benefits. Indeed, throughout the world, one finds people who grew up on farms, spent their working life outside of agriculture, and then become 'farmers' in retirement.

While 'corporate farms' have attracted much attention in the media, and it is true as we discuss in Chap. 5 that large-scale commercial farms produce the majority of agricultural commodities in industrialized countries, the fact is that most farms in the world are owned and managed by farm families (Lowder et al. 2016). A household is itself a complex human and social entity with many motivations, with the most basic one being to provide for the well-being of the family members. As the right-hand side of Fig. 3.1 illustrates, from an economic perspective, a household can be viewed as carrying out a number of 'household production' activities that involve the purchase of 'commodities' such as unprepared food and their transformation into 'goods' for consumption such as prepared meals. In the typical economic model of a household (originally developed by Becker 1992), the household makes these choices to achieve a particular objective, subject to the household's 'full income' which is

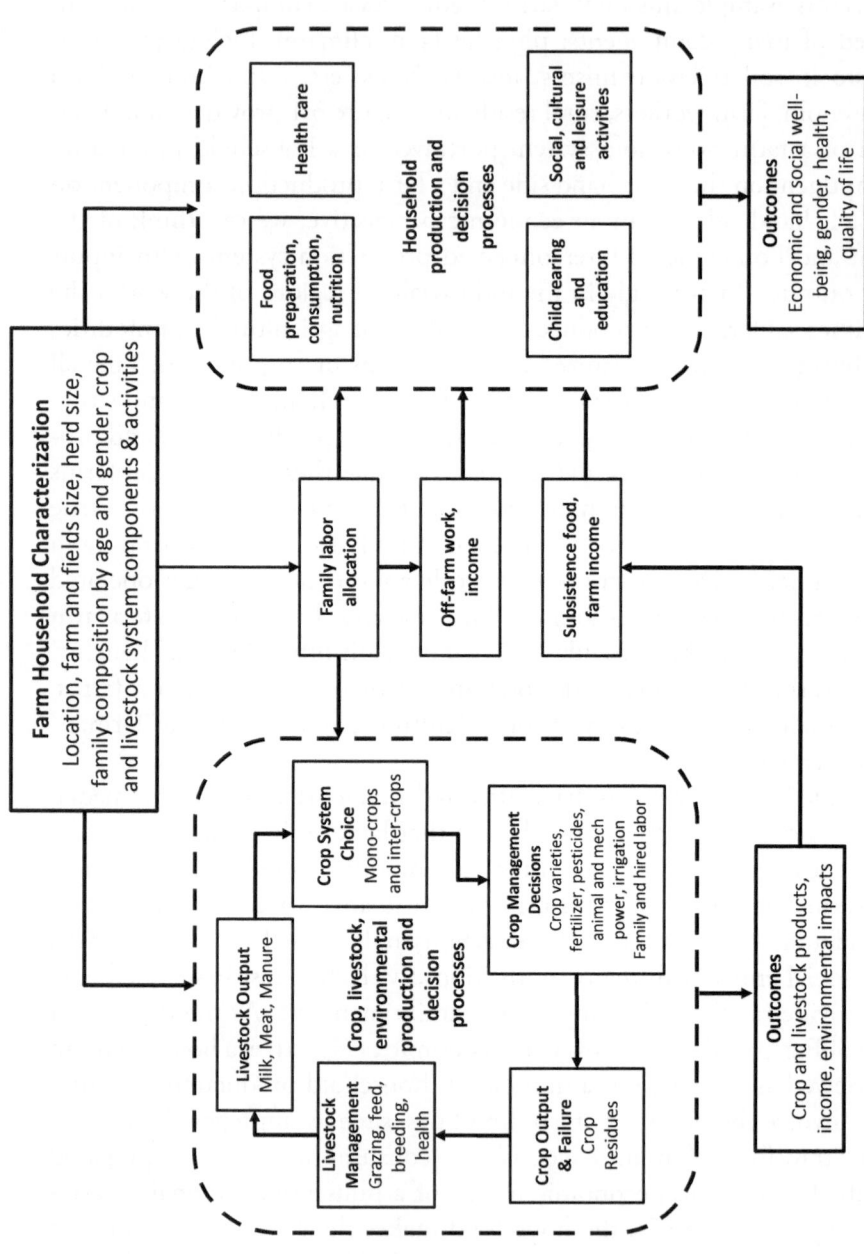

**Fig. 3.1** Farm household represented as farm and household production systems and decision processes. (Source: Jones et al. (2017))

made up of labor income, farm income, and other sources of income. An important element of Becker's model is the allocation of the household members' time between working outside the household to earn labor income, working in the household to produce consumption goods from purchased commodities, and other uses of time (referred to generally as 'leisure'). Other household activities can also be included in the household model as illustrated in Fig. 3.1, for example, providing child care.

An important feature of the farm household system in Fig. 3.1 is the interaction between farm and household components. For example, labor can be allocated to household activities or to farm production, as well as to working off-farm and to leisure. Additionally, the farm's production of food commodities can be used as inputs into household production along with purchased commodities. These linkages between the household's labor and commodity use creates what economists call a 'non-separable' farm household model, meaning that the production and consumption decisions of the household depend on farm production decisions such as crop choice and labor allocation. This non-separable feature is typical of farm households in developing countries where a substantial amount of farm production is consumed by the farm household, and also a substantial amount of household labor of both adults and children is allocated to farm production. In contrast, in typical farm households in commercial agricultural systems such as the United States, the farm and household components of the system are 'separable.' That is, the principal objective of the farm component is to produce income, and household food and other consumption decisions are made, given the income available from the farm business as well as from off-farm work and any other sources of income. In these commercial systems, one or more family members may work full or part time on the farm, but once that decision is made, the household production and farm production decisions are largely independent.

An issue that arises in understanding farm household systems is how farmers (or farm families or farm managers) make decisions, both about the farm and about other household activities. Decision-making within the household and the farm are typically represented in economic models as 'purposeful' or 'rational.' In the case of large commercial farms established by a group of investors, the goal is often explicitly defined as maximizing profit or return on investment. But as we noted above, if a farm household's well-being largely depends on the income generated by the farm, then it also must be able to operate profitably to remain in business and to meet the family's objectives. That being said, it is also important to understand that rationality does not imply that farmers only care about their money income. The farm household's objectives can be represented narrowly as maximizing the utility (or value) of net income from the farm but also can be represented more generally as depending on other factors such as managing financial risk, the nutrition and health of family members, or 'altruistic' objectives such as minimizing adverse impacts of their production activities on neighbors or the environment. Recent behavioral research suggests that 'bounded rationality' may describe the way people make decisions, due to limited information and the difficulties of choosing among a large number of options. It is important to keep in mind that rationality does not imply decision-makers do not make errors, nor does it imply that everyone would make the same choices in the same circumstances. Indeed, one implication of the heterogeneity of the physical and economic conditions in which farm households operate is that rational farmers should exhibit a wide range of behaviors.

## 3.3    Characteristics of Farm Household Systems

As we noted in the introduction to this chapter, a major challenge in sustainable agricultural development is the diversity and heterogeneity of farm household systems. To help deal with this challenge, scientists have developed a variety of methods to classify agricultural systems. Some focus on the physical characteristics of land and climate and the potential productivity of major crops (sometimes referred to as land suitability). Spatially referenced data, including remotely sensed land cover and land use data from satellites and aircraft, and from administrative data such as agricultural census and farm financial data, now provide the capability to characterize agricultural land use, crop and livestock production, environmental conditions, and socio-economic conditions. For example, the Food and Agriculture Organization of the United Nations (FAO) devel-

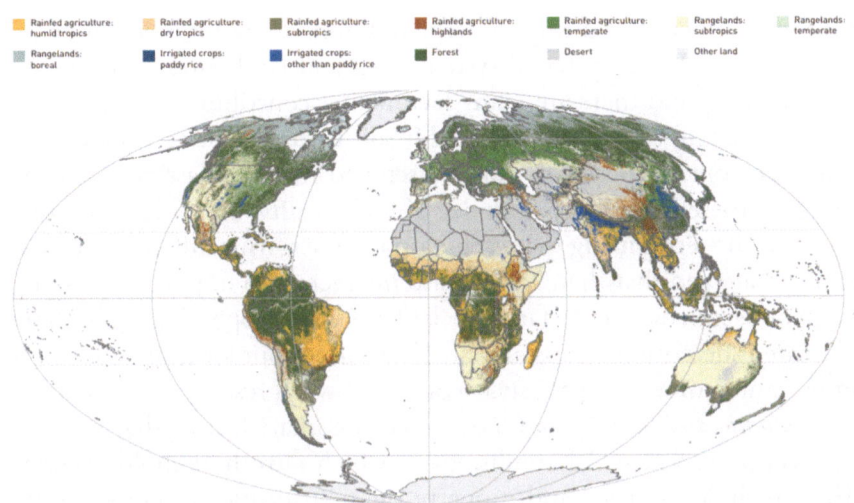

**Fig. 3.2** Global agro-ecological zones (GAEZs) map. (Source: FAO (2011))

oped a classification system that they use to define global agro-ecological zones (GAEZs), based on spatially referenced soils, climate, agricultural crop production, and other data (Fig. 3.2). In the United States, the US Department of Agriculture has created a hierarchy of Major Land Resource Areas based on soils, climate, and other bio-physical data. Recognizing the need to address more than the production component of agricultural systems, some more recent initiatives have added a socio-economic dimension to system characterization (Nemo 2019). Also, there are many activities to tailor classifications to specific regions and types of analysis (e.g., Andersen et al. 2006).

Because of the highly diverse and complex nature of farm household systems, there is no one correct way to describe them. For our purpose of understanding sustainable agricultural development, we shall follow the concepts embodied in Figs. 1.1 and 3.1. As Fig. 1.1 shows, farm household systems are linked to physical and human systems that define the conditions in which they operate, as well as the effects they have on other systems.

*Physical Environment: Climate, Soils, Topography, and Location.* Climate is basic to all biological systems, and the distinction used in the GAEZs between temperate and tropical, arid and humid environments is generally accepted. For crop-based systems, soil and land conditions are also fundamental, and there are many factors that contribute to soil productivity including soil type, depth, pH, soil organic matter content, and slope. For our purposes, it will be useful to refer to classifications that integrate these factors into an index of 'soil fertility' or 'inherent productivity,' while recognizing that each crop type performs differently to soils and climate and that these factors also interact. Also important is topography: it is much easier to farm on flat land than steeply sloping hillsides. Climate is determined substantially by location, but location also can be important in other respects, such as access to water resources and markets.

*Economic and Social Conditions.* There are many factors that make up the economic and social conditions in which farm households operate. These include the legal and property rights institutions, physical infrastructure, social attitudes, and various aspects of public policy. We often make use of the fact that most of these factors correlate strongly with basic economic development indicators, allowing us to use GDP per capita as a key proxy for social and economic conditions external to the farm household when comparing across countries, or per capita income when comparing across households within countries.

*Farm System.* We shall focus on the types of products that are produced and how they are produced. Figures 3.3 and 3.4 show where cropland is located globally, as well as the distribution of major crops across the globe, and Fig. 3.5 shows the global distribution of major livestock sys-

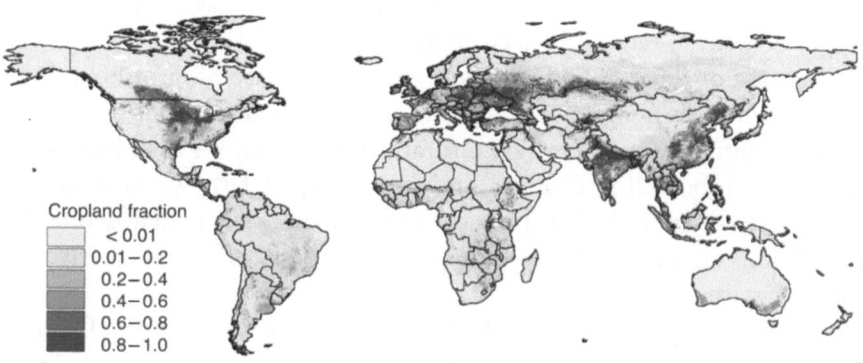

**Fig. 3.3** Global cropland distribution. (Source: Leff et al. (2004))

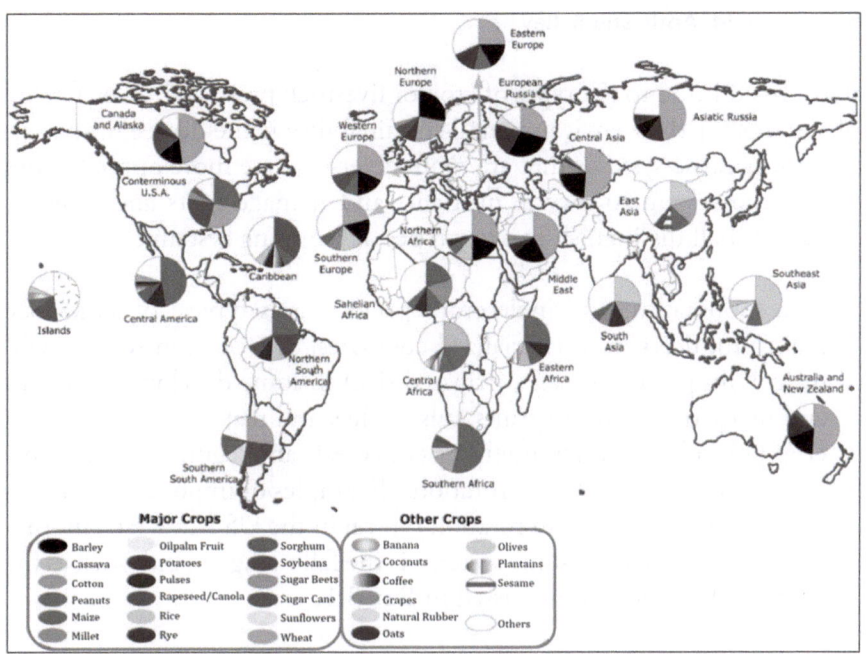

**Fig. 3.4** Geographic distribution of major crops across the world. (Leff et al. 2004)

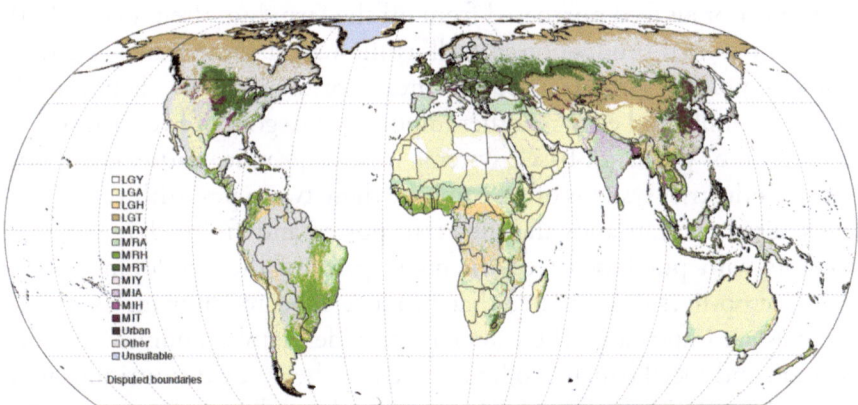

**Fig. 3.5** Global livestock systems. LGY: Hyperarid rangeland-based, LGA: Arid/semi-arid rangeland-based, LGH: Humid/subhumid rangeland-based, LGT: Temperate/tropical highlands rangeland-based, MRY: Hyperarid mixed rainfed, MRA: Arid/semi-arid mixed rainfed, MRH: Humid/subhumid mixed rainfed, MRT: Temperate/tropical highlands mixed rainfed, MIY: Hyperarid mixed irrigated, MIA: Arid/semi-arid mixed irrigated, MIH: Humid/subhumid mixed irrigated, MIT: Temperate/tropical highlands mixed irrigated. (Source: Robinson et al. (2011))

tems. In addition to the types of crop or livestock produced, a key feature of systems for their environmental sustainability is the degree of specialization, because systems that are more diverse create more opportunities within the system to recycle nutrients and manage pests and diseases. Thus, we shall distinguish two key types of cropping systems:

- crop monoculture (a single crop species), as is typical of large-scale grain producers in industrialized countries, as well as in some small-scale grain producers (e.g., irrigated rice) in some developing regions, and many tree crops for fruits, oils, coffee, and tea;
- crop polycultures, often intercrops of cereals and legumes, or multiple cereal and legume crops in rotations. Examples of important rotations include the corn-soybean rotation typical in the US Midwest. Another important element in some systems is the fallowing of land, such as the wheat-fallow rotation discussed in Box 3.1.

Livestock systems are complex because they involve the dynamics of animal reproduction and growth as well as managing animal behavior. Additionally, livestock systems may involve the production of the animals with most or all feed purchased from off the farm but often involve both feed and animal production as well as the use of some purchased feed. These types of systems produce meat (primarily beef, lamb, pork, chicken), milk, and eggs. Another important category of livestock system is pastoral or grassland-based systems used to produce cattle, goats, and sheep. Industrial systems often combine these two types, with young animals raised on grasslands and then transported and 'finished' in confined feedlots with purchased feed. In many parts of the developing world, mixed crop-livestock systems that produce food for humans as well as animals are important. These systems provide many economic, environmental, and social benefits to households, as they provide nutritious food for the household. They tend to be diversified and thus resilient to weather and other external disturbances, and they provide a way for farmers to maintain soil productivity by recycling nutrients between the crop and livestock components. This type of system is described in the Kenyan example of Box 3.2.

**Box 3.1   Commercial Wheat Farms in the US Pacific Northwest**

The inland Pacific Northwest of the United States is similar to other small grain-producing regions of the world, with a temperate semi-arid climate that supports a single growing season for wheat. Population density in rural areas is low, with small towns that range from a few hundred to a few thousand inhabitants that provide educational, financial, medical, and social services. The average household size is about 2.5 persons.

Commercial-scale dryland wheat production occurs on farms that average about 4000 acres (1600 hectares), but vary from 1600 acres to 7500 acres and produce about 85% of the wheat in this region; smaller farms average about 650 acres. A typical field, managed as a unit, is about 160 acres (60 hectares). These farms attain relatively high yields for dryland systems (averaging about 50 bushels per acre or 3.3 metric ton per hectare) on relatively fertile soils and intensive management. However, both yields and costs of production vary widely across the region with soil and climatic conditions, so that profit per acre also varies widely, from near zero to as high as US$100/acre in a typical year. In areas with low rainfall, winter wheat is grown in rotation with fallow as a way to manage soil moisture (winter wheat is planted in the fall and emerges, lies dormant over the winter, and then matures and is harvested the following summer; the next season the field is fallowed). In higher-rainfall areas, wheat is grown in an annual rotation with summer crops, often legumes. There are also smaller, less-specialized farms in the region producing wheat, other crops, and livestock. These are typically part-time operators who earn the majority of their income from non-agricultural activities.

Cropping operations are carried out with large-scale machinery for land preparation, planting, tillage, and harvest. Farm technology includes: commercially produced, improved seed varieties; chemical weed, insect and disease management; large four-wheel-drive diesel tractors with satellite-aided steering systems and computer-aided input application systems. Some farms also utilize large center-pivot irrigation with electrical or diesel pumps. Many farms use computers to access climatological and market data, and most have cell phone and Internet access. Due to the relatively light soils and hilly terrain, there is substantial wind and water erosion, and major efforts to develop and implement improved soil management practices have been made in the region. About half of the farms in the region employ reduced- or no-tillage methods of cultivation that require specialized implements for planting in fields that are not tilled or plowed between crops.

Commercial wheat farms are typically operated by a single household as a proprietorship or partnership with family members and may be incorporated, with the head of household and farm manager having between twelve and sixteen years of formal education. There are also large agribusiness firms, typically organized as corporations, owning or leasing land and hiring individuals (sometimes former farm owner-operators) to provide management as well as labor for machinery operation. Family members not engaged in farm operations often work off-farm, particularly those with relatively small operations.

## Box 3.2   Semi-Subsistence Crop-Livestock Systems in Machakos, Kenya

Machakos is a hilly drought-prone farming area centered about 50 kilometers southeast of Nairobi in Kenya, comprising the Machakos and Makueni districts. The agricultural systems of the region are typical of mixed crop-livestock systems in Eastern Africa. The Machakos study area illustrates the bio-physical and socio-economic heterogeneity typical in African systems. Altitudes range from 400 meters to 2100 meters above sea level in the semi-arid region with low soil fertility. Farm size varies highly between the villages with small farms (<2 hectares) in the higher, more humid areas and larger farms (about 8 hectares) in the lower and dryer rangelands. Typical fields are less than 0.5 hectare, with many fields less than 0.1 hectare and a small number of fields larger than 1 hectare. Per capita income varies greatly between villages, consequently poverty rates range from 37% in villages that produce cash crops such as fresh vegetables and milk for local or urban markets to 93% in areas farther from urban markets producing primarily maize and crops for household consumption. Average household size is about eight persons.

Farms include a variety of cropping systems in combination with livestock production. Semi-subsistence farms typically include monocrops, simple intercrops, and complex intercrops. In a simple intercrop, a small number of different species, such as beans and maize, are planted together. In a complex intercrop, a relatively large number of species are planted together, often in diverse combinations of crops in different proportions on different farms. Farms large enough to produce adequate feed typically own one to three dairy cows and have substantially higher incomes than households without livestock. In some areas, livestock is grazed, but many farms use a so-called zero-grazing system in which livestock is confined to increase the efficiency of nutrient recycling, using Napier grass and crop residues as feed and storing manure for use on the subsequent season's crops. Across the region, about one-third of farms produce milk, but the importance of dairy production varies greatly from one village to another. The villages with more cash crop production also produce much more milk, and dairy productivity varies substantially by village.

Productivity and the use of modern technology vary substantially across farms in this region and across Kenya. Maize yields average about 1.5 tons per hectare and are highly variable, with some farms obtaining less than 0.5 ton per hectare and some as high as 3 tons per hectare Crop failure rates are high for maize and beans in this semi-arid environment, as high as 50% in droughts, except for irrigated high-value crops such as vegetables. Crop byproduct is a small part of total crop value but is important for livestock feed and nutrient recycling, with about 50% of fields having manure and composted crop residues applied. Use of modern technology is limited. Only about 20% of maize parcels have fertilizer applied to them, and where it is used, the rate is about 65 kilograms of commercial product per hectare, a relatively low rate compared to the fertilizer recommendations. Commercially produced hybrid maize seeds are used in more productive highland areas but are used by few farms in more marginal lowland areas. Pesticide use is limited mostly to irrigated vegetables. Most fieldwork is carried out by hand labor, primarily family labor with limited use of hired labor. Some field preparation (plowing) is done with animals.

Another key element of how crops and livestock are produced is what we call the *production technology*. Important components of agricultural technologies include:

- improved genetics of crops and livestock for higher yields and other desirable traits such as drought and heat tolerance, nutritional characteristics, and visual appearance and taste;
- mechanical power and implements for crop management (land preparation, planting, cultivation, pest control), irrigation, harvest, and transport;
- chemicals for nutrient management and pest control;
- information and computer technology for data acquisition, production management, risk management, and marketing; and
- veterinary services including drugs for livestock disease management.

Two other features of farm systems that play an important role in their economic and environmental performance are *management ability* and *scale of production*. Management ability of farmers depends on their innate abilities but also importantly on their education and experience, particularly as the complexity of the technology and scale of operation increases, and utilization of information technology increases. Modern commercial farm operations require production expertise as well as business management expertise. Larger commercial operations typically employ individuals that specialize in different elements of both production and business management and often hire specialized services, for example, for nutrient and pest management, risk management and marketing, accounting and tax preparation, and regulatory compliance.

Closely related to management ability are the farm size and the scale of production. These features are closely related to the economic performance of the farm as well as the capability of the farm to provide income to the household. As we shall discuss in Chap. 4, for example, one of the factors causing farms in many parts of Africa and South Asia to be economically inefficient is their extremely small size. On the one hand, some of the technologies mentioned above, notably improved genetics and chemical technologies, may be used efficiently on very small farms as well as on much larger farms, because their use does not involve a large fixed capital investment. On the other hand, most mechanical power technologies (tractors, implements, harvesters, information technology) involve substantial capital investment and may only be useful on farms that exceed a minimal size and also may require higher education and specialized training. The obvious example that is widely used now in industrial systems is the use of large horsepower tractors with guidance systems linked to global positioning systems, but this is also true for the use of other kinds of information technologies, for example, to remotely sense crop growth and soil moisture to tailor applications of nutrients and irrigation water. Farm size and scale of production also can play an important role in purchasing inputs at competitive prices, in efficiently marketing products and managing risk, and in storage and transportation of commodities.

*Household System.* We focus on some key elements of the farm household, with somewhat different elements relevant to smaller-scale farms in developing countries and large, commercial farms in the industrialized countries. Family size and composition by age and gender play an important role in the developing world, as well as the type of work done by family members. The household's dependence on its own food production versus purchased food is a key feature. Attitudes toward investment in children's education and health, gender roles, social interactions, and non-farm income all can be important. The age and education of the family members also play an important role. In many parts of the world, off-farm work plays an important role in supplementing the farm household's income and in diversifying the farm household's economic activities.

## 3.4   Economic Analysis of Agricultural System Sustainability

Here and in later chapters, we use the examples presented in Boxes 3.1 and 3.2 as well as other examples to illustrate the challenges in moving agricultural systems in more sustainable directions. These examples illustrate one of the central themes of this book, namely that there are almost always tradeoffs between economic, environmental, and social dimensions of agricultural systems as we attempt to move them in more sustainable directions. In this section, we further explore the economic logic underlying these tradeoffs. Next, we discuss the conventional framework economists have developed for analyzing private and public choices among alternatives known as benefit-cost analysis (BCA). We explain some key limitations of BCA for analysis of sustainable development pathways and some advantages of using the tradeoff analysis approach that we introduced in Chap. 2 to evaluate agricultural system sustainability.

## 3.4.1  Economics of Tradeoffs in Agricultural Systems

How can we understand why tradeoffs are usually involved in moving agricultural systems in more sustainable directions? One way to think about this is to view each of the three dimensions of sustainability—economic, environmental, and social—from its own disciplinary perspective and then try to see how they interact as a more complex system. From an economic perspective, we can think of the farm household as making farm management and household decisions so as to achieve its objectives—let's just say the household's objective is to use the farm to earn income to support the family. That means that the farm is being managed to earn as much income as possible given the available resources, and following the wheat example of Box 3.1, suppose the farm produces wheat in a fallow rotation. Now, suppose we introduce the option of also producing the oilseed crop, camelina, to supply the biofuel industry, to sequester soil carbon, and reduce soil erosion and the net greenhouse gas emissions from the system. However, as we shall discuss in Chap. 6 (Box 6.5), at the likely market prices for wheat and camelina, the wheat-camelina system is likely to reduce wheat yields and the profitability of the system. Thus, farms adopting the wheat-camelina system would have to accept lower incomes in order to produce the additional environmental benefits associated with that system.

This example illustrates an economic principle that leads to tradeoffs being the rule rather than the exception in complex systems. If the system is 'optimized' in one dimension—for example, if the most profitable option for farmers in the dry rainfed regions of the US Pacific Northwest is the wheat-fallow rotation—then changing the system to increase its environmental sustainability, for example, by introducing camelina into the rotation, will necessarily reduce profitability. The only way that making a change will not be likely to reduce the economic performance of the system is if farmers are not managing for profitability, either because they have other objectives or because they are not capable of managing profitably. But there is strong evidence that most of these farms—that have been in the wheat business for many years—*do* know how to produce wheat profitably, and even casual conversations with farmers in the region indicate that they do, indeed, have profitability as their principal objective.

Now, suppose that rather than aiming for profitability alone, wheat farmers do in fact have other social objectives in mind. For example, many farmers may feel strongly about the role that US agriculture plays in helping to feed the world. Moreover, the US Navy, one of the potential customers for the biofuel produced from camelina, also aims to produce fuel domestically for national security and also without adversely impacting food production. That is, the Navy does not want to induce a *tradeoff* between global food needs and national security by using biofuels. But, as the example shows, introducing camelina into the wheat-fallow system is likely to reduce wheat production, by both reducing wheat yields when wheat is grown and the amount of acres in wheat cultivation each growing season.

Finally, consider the effects of introducing camelina into the system from the farm household's perspective. The farm family may have an objective of being involved in other economic or leisure activities that are made possible, in part, by the wheat-fallow system that involves having one-half of the farm's land under cultivation each season. Introducing camelina into the system would mean cultivating as much as 100% of the farm's land each season, thus requiring more labor hours to manage the farm and also to perform operations such as land preparation, planting, and harvest. So either more family labor time would have to be devoted to the farm or additional labor would have to be hired and managed. Thus, in order to adopt the wheat-camelina system, the farm household could face tradeoffs between its professional and lifestyle objectives and the economic and environmental performance of the farm production system.

## 3.4.2 Benefit-Cost Analysis and Its Limitations

Benefit-cost analysis (BCA) is the standard economic framework for analysis of both private economic decision-making, such as the farm management decisions discussed in the previous section, and analysis of public policy issues such as how to regulate pollution. In its simplest form, BCA of a production system is equivalent to a calculation of net economic benefits or profitability either in a single-period (static) setting or more generally over a relevant time horizon in which a change in technology or policy has effects on a system. When only 'private' benefits and costs are involved, BCA is equivalent to standard economic analysis of an investment in which the net present value of benefits and costs of the alternative technology or policy is compared to the current or 'baseline' case. BCA is also used to analyze multidimensional outcomes involving both market and non-market changes, as when a system is producing a product like wheat but also causing an environmental externality such as water pollution caused by soil erosion. In these analyses, market-based impacts (e.g., farm income generated by producing and selling crops and livestock) are combined with the value of 'non-market' outcomes such as changes in water quality. To account for the time dimension, the analyst assumes a 'social rate of discount' and calculates the change in the present discounted value of net benefits (benefits minus costs), both the 'private' net benefits (say, the wheat farmers' profits) and the 'social' impacts (say, the value of the damages caused by water pollution). In the simple case where the current technology or policy is compared to a single alternative, an alternative that yields an increase in net benefits (or equivalently, a benefit/cost ratio greater than 1) is judged to be preferred to the current one. In principle, if all policy options could be evaluated in this way, the best option could be identified. To implement this benefit-cost framework, credible estimates of quantities and values of marketed goods are needed (e.g., quantity and price of corn produced and cost of production), as well as quantities and values of non-market outputs (e.g., nutrient concentration in surface and groundwater and the environmental or health damages caused by it).

As this example illustrates, there are many conceptual as well as practical limitations to the BCA approach. One important issue is the effect of aggregating benefits and costs when they are experienced by different individuals. Early in its development and use, it was recognized that, by aggregating benefits and costs across gainers and losers, the distributional impacts of a change in policy or technology were being ignored in standard BCA. For example, a farmer may profit from the use of nutrients to produce corn in the US Midwest, but the shrimp industry in the Gulf of Mexico may lose from the impacts of hypoxia on shrimp populations in the Gulf. Aggregating benefits and costs into a 'net benefit' does not allow a decision-maker to know who benefitted and who lost, or by how much.

Another key problem with BCA is the valuation of benefits and costs. While it is straightforward to measure and value of market outcomes such as the amount and value of corn produced, and the loss of shrimp produced, it is difficult to quantify and value non-market outcomes such as the human health effects of nutrient pollution of drinking water. With adequate scientific understanding, relevant data and suitable measurement technologies, it is possible to objectively quantify the non-market outcomes. But, in many cases, valuing non-market outputs is exceedingly difficult. For example, contamination of water by nutrients such as nitrates may have adverse impacts on human health, and it may be possible to estimate the magnitude of these effects, but it is difficult to attach a monetary value to health effects that is generally accepted by the affected people or by society as determined through political processes. Similarly, ecosystem services such as biodiversity are difficult to quantify and value in monetary terms. Moreover, many people may object, on ethical grounds, to the idea that the value of human health or the environment can be monetized.

Another limitation of BCA is its lack of transparency across the various dimensions of impact, a more general aspect of the aggregation problem mentioned above. When benefits and costs involve multiple dimensions, both market-based and non-market such as the example above, summarizing them in a single metric such as a net benefit or benefit-cost ratio does not convey information about the changes in the individual dimensions. Different stakeholders are likely to attach different degrees of importance or values to the various dimensions of impact—for example, some people may care more about farmers' incomes, others may care more about water quality and health. Attempting to attach a monetary value to each outcome and then aggregating them obscures both the impacts that a technology or policy may have and the values used to aggregate across dimensions.

### 3.4.3 Tradeoff Analysis as a Participatory Process

We described the tradeoff analysis approach using aggregate indicators, such as those associated with the sustainable development goals, in Sect. 2.7 of Chap. 2. A challenge with setting goals at that level is ensuring participation and buy-in by the people who are both impacted by the goals and who must undertake actions to implement change. As we discuss in Chap. 6, perhaps the most daunting challenge of moving agriculture in more sustainable directions is in the implementation of needed changes to achieve sustainable development goals. At the local to regional level where agricultural systems operate, this connection between the people who are impacted by sustainable development, and the process of designing and implementing sustainable development pathways, is much more feasible.

Figure 3.6 provides an example of how a participatory approach to tradeoff analysis, as the tool used to implement the positive dimension of sustainable development, can be implemented. The top box in Fig. 3.6 focuses on stakeholder-led processes whereby the decision context and goals and potential solution sets are identified through stakeholder engagement with experts, local communities, institutions, and scientists. Outputs of this process are then fed into the lower box representing the technical, analytical component of tradeoff analysis. The analytical component translates stakeholder preferences into quantifiable indicators, identification of appropriate analytical tools (data and models), and the implementation of analysis. The results of the modeling are then transformed into analysis and visualizations that are communicated to stakeholders for review. In most cases, this is an iterative process with multiple rounds of scenario co-production, analysis and review via stakeholder consultation until realistic scenarios and tradeoff results are found.

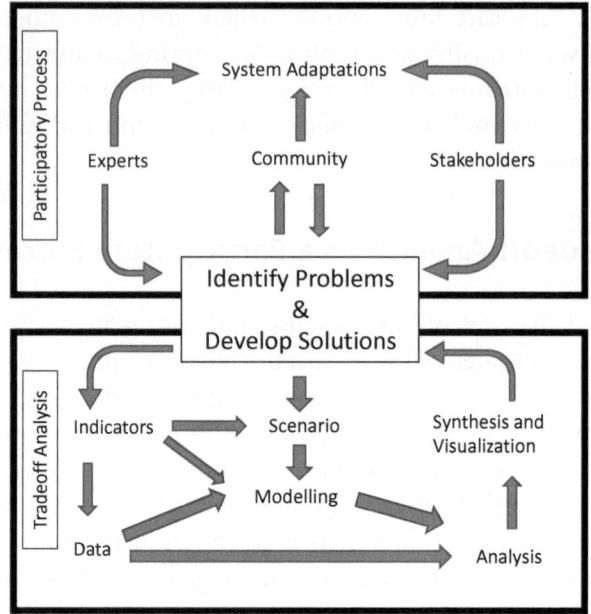

**Fig. 3.6** Participatory modeling using tradeoff analysis. (Source: Based on Kanter et al. (2018))

## 3.5 Farm Household System Sustainability Indicators

A key part of tradeoff analysis is the identification of sustainability indicators that can be used to set goals and to evaluate and compare the performance of systems along alternative development pathways. Chapter 2 discussed the sustainable development goals and the indicators that are used to quantify development pathways at the national level. At the farm level, an analogous set of economic, environmental, and social indicators have been identified and used. Some of the widely used indicators are listed in Tables 3.1, 3.2 and 3.3.

**Table 3.1** Economic indicators and SDGs

| | Economic Indicators | Unit | SDG |
|---|---|---|---|
| Farm | Crop or livestock productivity (yield) | quantity/ha or quantity/animal | SDG1: No poverty |
| | Financial condition | Debts/assets (%) | SDG2: Zero hunger |
| | Farm income | Currency units per farm, per ha, per animal unit | SDG8: Promote sustained inclusive and sustainable economic growth |
| | Technology: improved genetics; purchased inputs (seeds, fertilizers, pesticides); mechanical power and implements; information technology | use/non-use; application rates (quantity/ha) | SDG13: Climate action |
| | Diversification or resilience | Crop or livestock species diversity index, drought or disease tolerance | |
| Household | Money income | All farm and non-farm money income (currency units/time) | |
| | Full income | Net value of all farm and household production and labor plus non-farm money income (currency units/time) | |
| | Poverty | % individuals or households with consumption or income below poverty line | |
| | Vulnerability | Possibility of suffering a decline in well-being due to an adverse shock | |

**Table 3.2** Environmental indicators and SDGs

| | Indicator | Unit | SDG |
|---|---|---|---|
| Soil | Soil organic matter | % of soil | SDG2: Zero hunger, SDG6: Ensure availability and sustainable management of water and sanitation SDG12: Responsible consumption and production SDG15: Halt biodiversity loss |
| | Soil fertility | Ph, macro-micro nutrient balance | |
| | Soil erosion | kg/ha | |
| Water | Depth of groundwater | m | |
| | Water quality | Ph, salinity | |
| | Dissolved oxygen in water | %, mg pollutant/l, ppm | |
| | Heavy metal concentration in water | %, mg pollutant/l, ppm | |
| Emissions | Carbon dioxide ($CO_2$), methane ($NH_4$), nitrous oxide ($N_2O$) | kg GHG/year | |
| | Air pollution | Air quality indices (AQI) | |
| Land Use | Increase in forest cover | Share of land converted to protected forests, share of degraded land recovered | |
| | Conservation of fragile ecosystems | Conserved ecosystems in coastal, mountain, and island systems | |
| Biodiversity | Species richness | Gamma diversity: count of species in a region | |
| | Protection for terrestrial and freshwater biodiversity | Share of land protected | |

**Table 3.3** Social indicators for human development, equity and justice, and SDGs

| | Indicator | Unit | SDG |
|---|---|---|---|
| Food Security | Availability of culturally relevant food | Calories/capita, expenditure/capita | SDG2: Zero hunger |
| | Access to food | Household Food Insecurity Access Scale (HFIAS), Food Insecurity Access Scale (FIES) | |
| | Diet diversity | Food Consumption Score (FCS), Household Diet Diversity Scale (HDDS) | |
| | Food safety | % population with access to clean drinking water, number of food contamination cases due to pathogens and agri-chemicals; cases of food-borne illness or death | |
| Nutrition | Undernutrition | % of children stunted, wasted, or underweight, % with micronutrient deficiency | |
| | Overnutrition | % obese, % calories from saturated fats, | |
| Health | Maternal child mortality | Maternal mortality ratio, infant mortality rate | SDG3: Good health and well-being |
| | Mental health | Cases of illness or deaths | |
| Women's Empowerment | Women's land ownership, control over assets, and autonomy in the household | % of land owned by women, % of assets controlled by women | SDG5: Gender equality |
| | Pay gap | Relative difference in wages of men and women | |
| | Intra-household resource allocation to female members | Difference in food allocation, and educational investments between girls and boys | |
| | Women's Empowerment in Agriculture Index (WEAI) | Index | |
| Safe Working Conditions | Worker safety | Number of injuries or deaths | SDG8: Decent work and economic growth |
| | Exposure to harmful chemicals | % of workers without protective gear, % of workers exposed to harmful chemicals | |
| Community | Viability | Age distribution | SDG11: Sustainable cities and communities |
| | | Educational, medical, and social services | |
| Animal Welfare | Confinement practices | % confined | |
| | Animal health | disease and mortality rates | |

## 3.5.1   Economic Indicators

One of the foremost indicators for an agricultural system is its average productivity, typically defined as the amount of output per the amount of some input used to produce it, per unit time. For crops, the typical productivity indicator is crop yield defined as, for example, kilograms per hectare per growing season, or for livestock, the kilograms or liters of milk or meat per animal unit per time period. In some fields of science, average productivity is also called 'input use efficiency.' Actual yields obtained on farms depend on the physical and biological conditions on farms and on how they are managed.

Another way to measure productivity is to use experiments carried out under controlled management conditions which can estimate the 'potential' yield for the local climate and soil conditions. Typically, yields under experimental conditions are substantially higher than those obtained on farm fields. The difference between 'potential' yield achieved under controlled experimental conditions and actual yields observed on farmers' fields is called the 'yield gap' by agronomists. This yield gap is often interpreted as a measure of farm 'technically inefficiency.' This type of efficiency indicator is particularly relevant in developing countries where yield gaps are large for major food crops, and productivity continues to fall far below potential productivity. Under these conditions, adoption of improved farm management practices can increase yields while maintaining or improving the productive capacity of the systems. In industrialized agriculture where yield gaps are typically relatively small, sustainable development is oriented toward maintaining productivity near its potential while improving performance in environmental and social dimensions.

Other measures of productivity are used by economists in addition to crop or livestock yield. For example, output per unit of labor input or any other input can be calculated. However, it is important to recognize that all of these average productivity measures are 'partial' in the sense that they do not account for differences in other inputs. For example, when comparing maize yields on two farms, one with a low yield and one with a higher yield, the differences could be due to the use of other inputs such as fertilizer, irrigation, labor, mechanical power, and the ability of the farm manager. In an attempt to control for the amounts of all management inputs, a broader measure of productivity known as 'total factor productivity' is often used by economists. Total factor productivity is the ratio of output to an index of all inputs into the system. Typically, total factor productivity is used to measure the changes in productivity over time (rates of productivity growth or change) at the sector level, in which case the numerator is an index of all outputs produced and the denominator is an index of all inputs employed. These indexes represent an attempt by economists to aggregate across the various diverse and heterogeneous agricultural systems to obtain an aggregate picture of the productivity changes taking place in the agricultural sector. It is important to recognize that while total factor productivity has the advantage of being more comprehensive, it has the disadvantage of attempting literally 'to compare apples and oranges' and thus must also be interpreted with care. For this reason, at the farm level, the simpler partial productivity measures, such as individual crop yield, are more commonly used. To obtain a deeper understanding of factors influencing these partial productivity measures, economists often use observations of actual farm outputs and inputs to estimate statistical 'production function' models that explain changes in crop yield or livestock productivity across farms or over time in terms of all of the inputs used, as well as other observable characteristics such as soil type and climate, so that the contribution of each individual input to productivity can be assessed.

Like people everywhere, farm households are interested in increasing their incomes, which for farm businesses means trying to maximize the profit from the agricultural system, that is, the difference between gross value of production and the cost of inputs. Inputs include nutrients sourced via fertilizers, chemical treatments such as pesticides and herbicides, water for irrigation, energy to power machinery, and labor engaged in a host of activities ranging from on-farm cultivation to management. In calculating profit, an important distinction economists make is between 'accounting profit' or revenue net of cash operating costs and the 'economic profit,' which incorporates the operating costs as well as the implicit value of the farm owner's assets used in production, including the land owned by the farm and the value of the farm operator's time. We can think of a farmer that owns the land being operated as paying an implicit rental on the land equal to what the farm could earn by renting the land to someone else. The same principle applies to the machinery the farm owns and the value of the farmer's own time—both have a 'rental value' or opportunity cost equal to what they could earn elsewhere.

Accounting for the contribution of the farm's own resources is critically important to understanding a farm's 'real income' and the long-term economic sustainability of farms. For example, the increasing size of farms in industrialized countries (see Chap. 5) can be explained by the fact that smaller farms do not provide incomes comparable to non-agricultural occupations in the industrialized countries. Likewise, it is important to recognize that a large part of the cost of production for many small farms in the developing world is family labor. Even though this labor is not paid for by the family, it has an implicit value or 'opportunity cost'—that is, family members' time could be used to work off-farm or to do other work in the household such as food preparation or child care.

A widespread social concern is not only the average income in society but also the distribution of income. A number of measures of income distribution have been developed, but the most widely used indicator relates to poverty, typically measured as the share of a population earning an income per person (per capita) or per household less than a 'poverty line.' Most countries define their official poverty lines in relation to local living standards, often based on how much income is needed to purchase minimal food and shelter and other necessities. Also international organizations such as the World Bank define and report poverty rates based on a particular international standard. For example, for many years, extreme poverty was defined as a per capita income less than one US dollar per day, where the value was calculated using prices and incomes in a specified year. It is important to recognize, however, that these international poverty statistics are affected by the way that national currencies are valued and also are affected by changes in prices and incomes over time.

Economic indicators are highlighted in a number of sustainable development goals, including SDG 2, 'No Poverty,' and SDG8, 'Sustained Inclusive and Sustainable Economic Growth.' Other important dimensions of well-being related to economics—but difficult to quantify—are vulnerability and resilience. For example, even though poor households may be above a poverty line, they may also be affected by income shocks that could put them at risk of having their income fall below the poverty line, or at risk of food insecurity. Thus, diversification of income sources, such as having both farm and non-farm income, is one of the ways to increase resilience. This can also be measured in terms of diversity indices for sources of income within the agricultural sector as well as outside the non-farm sector.

## 3.5.2  Environmental Indicators

These indicators can be divided into four broad aspects of the environment—soil, water, air, and biodiversity. First, maintaining the quality and productive capacity of soil is essential for sustainable production of food for a growing population. Organic matter in the soil is usually derived from decomposition of plant residue and so depends on microorganism populations in the soil. Organic matter improves water holding capacity and is a source of soil organic carbon and the principal nutrients (nitrogen, phosphorous, and potassium) essential to plant growth. Crop growth utilizes these nutrients through photosynthesis to make plant biomass and must be replenished to maintain soil productivity. Another key factor affecting plant growth is the soil acidity or alkalinity (pH) which affects the ability of plants to utilize soil nutrients. The presence of toxic metals such as aluminum and minerals such as salt can also adversely affect plant growth. Agricultural production requires slightly acidic soils in general with varying requirements for optimal production of different crops. Soil erosion particularly for cultivation in steep slopes can lead to the loss of fertile top soil.

Water quality measures include its pH, salinity, and dissolved oxygen. Slightly acidic water is ideal for maximum nutrient uptake by plants. High percentage of soluble salts can hinder nutrient uptake as well as hinder growth. Nutrients lost from agricultural systems often accumulate in groundwater or drain into water bodies. Overuse of chemical fertilizer, pesticides, herbicides, and other synthetic inputs has led to severe environmental impacts such as hypoxia in many coastal waters around the world, particularly where large rivers transport water from intensive agricultural areas, as is the case where the Mississippi River flows into the Gulf of Mexico. Heavy metal concentrations in water bodies or low dissolved oxygen are indicators of poor water quality. Increasing depth to groundwater occurs with pumping of groundwater faster than it is recharged by infiltration.

Agricultural practices can also cause air pollution. Stubble burning in northwest India has been associated with spikes in air pollution in neighboring cities, and bare fallow fields in the US. wheat belt often cause dust storms that pollute air in nearby cities. Air quality indices are used to quantify such air pollution. Incidence of air quality-led respiratory diseases can also be used as a measure particularly relevant for developing countries where the average household is too poor to afford air purifiers. Agriculture emits the three principal greenhouse gases (GHGs)—carbon dioxide, methane, and nitrous oxide—through a variety of mechanisms. Conversion of land from forest or grasslands to agriculture is a major source of carbon dioxide, along with conventional cropland tillage and use of fossil fuels to produce and transport agricultural products. Use of nitrogen fertilizer is a major source of nitrous oxide. Flooded rice fields and livestock are major sources of methane.

Ecologists use measures of species diversity to indicate ecosystem health. Biodiversity can be measured as a count of the number of species in a region and also in soils as a measure of soil health. In agricultural systems, biodiversity can be increased through intercropping, planting trees or cover crops, and by practices such as hedgerows that provide wildlife habitat.

### 3.5.3    Social Indicators

In addition to economic and environmental indicators, sustainable development goals also include aspects of human development, equity, and justice which are measured by indicators discussed in Table 3.3. Food security is multidimensional in nature and best measured with a combination of indicators. A typical household's diet is comprised of various food items such as cereals, pulses, fruits, vegetables, dairy and eggs, meat, edible oils, sugar, and beverages. Total food consumption of a household can be measured in terms of calories consumed per capita, and the food security status is determined in reference to a benchmark minimum calorie requirement. In the absence of data on physical quantities of consumption, value of food consumed can be measured in terms of per capita expenditure on food. When detailed consumption or expenditure data are lacking for households, simpler proxies can be used, such as an income-based food security indicator based on the minimum income necessary for a nutritionally adequate diet (Antle et al. 2015a).

Measures of food availability or food expenditure do not capture important aspects of access, diet quality, and food safety. The Economic Research Service (ERS) of the United States Department of Agriculture developed the Household Food Security Survey Module (HFSSM) that records information on objective and subjective aspects of food security such as whether a household worried about meeting basic food requirements; whether they had to reduce quantity, quality, or frequency of consumption; and whether they felt hungry. The Household Food Insecurity Access Scale (HFIAS) and Food Insecurity Experience Scale (FIES) have been modeled on the HFSSM and used to monitor access to food insecurity in developing countries (Coates et al. 2006; Ballard et al. 2013). The FIES has been adopted by the United Nations and FAO to track progress in food security across the world under the SDGs. Diversity in diet is a good measure of diet quality since more balanced diets comprise macro- and micronutrients leading to better nutrient adequacy. The Food Consumption Score (FCS) is an index developed by the World Food Programme (WFP) that weights the frequency of food consumption in eight different food groups by their nutritional value to measure the diversity in diet. A simpler measure of diet diversity is the Household Diet Diversity Scale (HDDS) which is a count of the number of food groups consumed by a household. Another aspect of diet quality includes food safety. Measures of food safety include access to clean drinking water, cases of food-borne diseases, incidences of food contamination, and illnesses or deaths associated with these.

Closely related to food security are nutritional outcomes. Inadequate dietary nutrient intake and poor quality of nutrition weakens the immune system, increases susceptibility to diseases, and inhibits growth particularly among children. Undernutrition among children is measured by anthropometric indicators such as stunting (low height-for-age), wasting (low weight-for-height), and underweight (low weight-for-age). Stunting is a result of prolonged undernutrition, wasting results from inadequate nutrition for shorter periods and underweight is a combination of the two.

The lack of balanced diets with disproportionate focus on macronutrients can lead to micronutrient deficiencies, notably for vitamin A, iron, and iodine. These deficiencies among women in reproductive ages can adversely affect pregnancy outcomes. Infant and maternal mortality rates track the number of death of infants and mothers during childbirth.

The rising incidence of obesity is now considered a major global health problem associated with risks of diabetes and heart disease in the industrial as well as developing countries. According to the World Health Organization (2018), overweight and obesity are now linked to more deaths worldwide than underweight. Globally, there are more people who are obese than underweight—this occurs in every region except parts of sub-Saharan Africa and Asia. Overweight and obesity are increasing rapidly in low- and middle-income countries, particularly in urban settings. The number of overweight children under five years has increased by nearly 50% since 2000 in Africa, and almost half of the children under five years who were overweight or obese in 2016 lived in Asia. Although the obesity epidemic is not directly due to agricultural production systems, it is clear that the evolution of the global food system toward provision of low cost, calorie dense processed foods is playing a large role, along with the more sedentary lifestyle associated with industrialization and urbanization. As we discuss further in Chaps. 5 and 6, consumers' preferences for healthier and more sustainable diets can play an important role in moving the food industry, including agricultural production systems, toward supplying more affordable, healthier sustainably produced diets.

Measures of overnutrition such as percentage of obese is usually measured using the body mass index (BMI) calculated as a person's weight in kilograms divided by the square of the person's height in meters ($kg/m^2$). BMI provides a useful measure of overweight and obesity because it is easy to calculate but should be considered a rough guide because it may not correspond to the same degree of fatness in different individuals. Generally accepted standards for adults are: overweight is a BMI greater than or equal to twenty-five; and obesity is a BMI greater than or equal to thirty. For children, age needs to be considered when defining overweight and obesity (World Health Organization 2018).

As discussed in Chap. 2, women play a crucial role in the agricultural sector, particularly in developing countries, and an improvement in their status can have far-reaching impact for several SDGs. Additionally, reducing gender-based inequality is intrinsically valuable. The Women's Empowerment in Agriculture Index (WEAI) has become widely used. The WEAI is based on five domains: (1) decisions about agricultural production; (2) access to and decision-making power over productive resources; (3) control over use of income; (4) leadership in the community; and (5) time use decisions (Alkire et al. 2013). While the WEAI combines multiple aspects of women's empowerment, individual elements such as land controlled by women, pay gap, their participation in household decisions, and equality in intra-household resource allocation have been used to monitor the women's status. While the importance of these measures is widely acknowledged, reliable data sources on these indicators continue to be limited.

Well-being and safety of people working agriculture is another important social issue. Agriculture is one of the riskier occupations, due to exposure to dangerous machinery and chemicals. In the tropical developing world, exposure to a number of serious diseases is associated with farm work, particularly in humid environments where vector-borne diseases such as malaria and bilharzia are prevalent. Migrant workers in agriculture may be exposed to risks from racial discrimination and violence, unhealthy living conditions, and emotional stress. Mental health is also a serious concern for farm owners, with relatively high rates of family violence and suicide reported in both low- and high-income countries. Often, increased rates of violence and suicide are reported during times of extreme economic stress.

The viability of rural communities is another important social concern that can be strongly influenced by agricultural development. In both developing and industrial societies, out-migration of young people leaves rural populations with a disproportionate number of older people. This is a particularly serious problem in sparsely populated areas where it can reduce the economic base needed to support essential educational, medical, and social services.

We also include animal welfare in the social category, although it cuts across economic and environmental dimensions as well. As a social issue, we refer to the ethical concerns people have for the pain and suffering that farm animals may be subjected to as they are raised, used for production of dairy products and eggs, or slaughtered for meat. How animals are treated also can relate directly to the environmental dimension, for example, beef cattle that graze in pastures up to the time they are slaughtered may have much lower environmental impact than cattle confined in feed lots and fed grains and treated with antibiotics to control diseases caused by confinement. Likewise, these alternative production processes may have very different economic implications for farm incomes as well as the price of meat for consumers and also for the health of consumers and the general public.

## 3.6    Methods, Data, and Models to Assess Agricultural System Sustainability

A variety of quantitative tools are used to study tradeoffs in agricultural systems. The key analytical challenge is to predict changes in the sustainability indicators used to represent the performance of farm household systems. A major methodological challenge is to evaluate the performance of these systems when they are modified in some way, for example, when a new crop or crop variety is introduced into the system, or the farm household chooses to work less on the farm and more in other activities.

Evaluating the performance of farm household systems is challenging for several reasons. First, agricultural system complexity means that it is difficult—if not impossible—to use conventional experimental methods (often referred to as randomized controlled trials [RCTs]) to study farm household systems because the same farms cannot be observed doing two or more different things at the same time. Whereas an agronomist can carry out a yield trial for a new crop and make side-by-side comparisons of, say, an old variety and a new variety, it is very costly and fraught with other practical and ethical difficulties to do this with actual farm households (Barrett and Carter 2010). For example, when farmers adopt a new crop variety, they may make a number of changes in related management practices, the farm household might change its food consumption behavior, and so on. Thus, it is largely impractical to use the standard experimental design methods to carry out RCTs to study the sustainability of farm household systems, although some experimental data, such as agronomic trials, and also some behavioral experiments, can provide useful information. Another reason why conventional field experiments are often not used to study system sustainability is because such trials are extremely slow, because they can only proceed as fast as crops and livestock growth, and as fast as household behavior can be observed under varying conditions.

Another major factor that limits the usefulness of RCTs to study agricultural system sustainability is that many of the important questions are about future, or prospective, systems, under future conditions not observable in historical data or in current experiments. The most obvious example is the study of how current or possible future agricultural systems will perform under future climate and socio-economic conditions.

These considerations have led many researchers studying farm household sustainability to use computer simulation models to mimic the structure of the farm household such as the one illustrated in Fig. 3.1. These computer models are used to carry out simulation experiments to project changes in the economic, environmental, and social indicators described in Sect. 3.5. Various measures of farm household well-being are used, such as farm income and its distribution among geographic regions and among different types of farms. Measures of environmental outcomes and ecosystem services are available from direct measurements and from models, including soil quality and productivity, air and water quantity and quality, greenhouse gas emissions, and wildlife habitat. For example, the US Department of Agriculture has constructed an 'environmental benefits index' to assist in the design and implementation of conservation programs that combines a number of different environmental indicators such as those in Table 3.2 into a summary measure. Models are also available to quantify food security indicators and some health-related indicators, but as yet models do not exist for other social outcomes, and analysts must utilize qualitative assessments in conjunction with quantitative analysis of the quantifiable economic and environmental dimensions.

## 3.6.1 Regional Integrated Assessment Modeling Methods

We illustrate the use of agricultural systems models to assess farm household sustainability and tradeoffs with the regional integrated assessment (RIA) method developed by the Agricultural Model Inter-comparison and Improvement Project (AgMIP) (Antle et al. 2015b). The approach is designed to quantify indicators of system performance deemed to be relevant by both stakeholders and scientists and then conduct simulation experiments to evaluate how system performance responds to climate and adaptive responses to climate change, similar to the process illustrated in Fig. 3.6. The RIA approach can be linked to global modeling that provides inputs, such as global market prices, into the regional analysis, as illustrated in Fig. 3.7. These methods can be used in various ways to support analysis of farm household system sustainability, for example, to facilitate the targeting of agricultural interventions to farm types, for design and impact assessment of context-specific safety net, food security, or market-oriented intervention packages.

The foundation of the RIA approach is the design of the simulation experiments that are used to evaluate climate impacts and the effects of system adaptations. There are many possible simulation experiments that can be carried out. Here, we illustrate how the RIA approach is used for analysis of sustainable development pathways in response to climate change, as elaborated in Valdivia et al. (2015). This approach to regional impact assessment uses four 'core' research questions as the basis for simulation experiments:

*Question 1*: What is the sensitivity of current agricultural production systems to climate change? This is the analysis of the 'business-as-usual' system that can be compared with alternative systems, for example, ones better adapted to changes in climate and designed to be more sustainable in relevant dimensions.

*Question 2*: What are the effects of adaptation in the current state of the world? This question is one often raised by stakeholders—what is the value of adapting today's agricultural systems to climate changes that may be occurring now and in the near future, assuming that most other social and economic conditions are similar to today?

*Question 3*: What is the impact of climate change on future agricultural production systems? This question evaluates the isolated role of climate impacts on a future production system, which will differ from the current production system due to development in the agricultural sector not directly motivated by climate changes. For example, this could be a system designed to be more economically and environmentally sustainable and motivated, for example, by the goal of reducing water pollution, but it is not intended to necessarily better adapt to a changing climate.

*Question 4*: What are the benefits of climate change adaptations? This question analyzes the benefit of potential adaptation options in the production system of the future, which may offset climate vulnerabilities or enhance positive effects identified in Question 3 above.

In the AgMIP RIA methodology, the heterogeneous response of farms to climate change derives from the productivity impacts of climate change incorporated through crop and livestock simulation models, as well as the socio-economic heterogeneity in the farm household system represented in economic models due to variations in farm size, household size, and non-farm income. These elements operate at the global scale through the worldwide food system, as well as at national and subnational (or regional) scales (Fig. 3.7). For analysis of adaptations, a similar method is used to

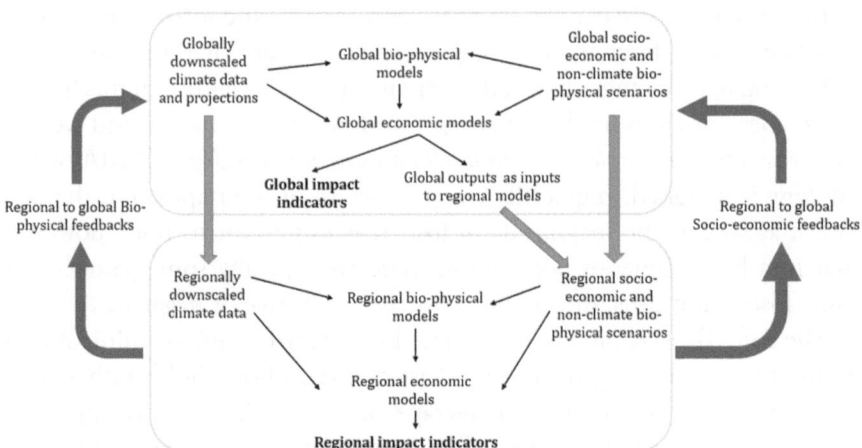

**Fig. 3.7** AgMIP global and regional integrated assessment (RIA). (Source: Based on Antle et al. (2017b))

assess how the existing system could be changed, by modifying the properties of the models. These changes can range from management of the existing production activities, changes in the land or other resources allocated to those activities, as well as the introduction of new activities or the elimination of activities. These characterizations of the existing and prospective farming systems also help to develop future socio-economic pathways (i.e., Representative Agricultural Pathways discussed in Chap. 6) by identifying the 'external' or 'driving' variables that define the bio-physical and socio-economic conditions in which the analysis is conducted. For example, if the analysis is being designed for a future period, it is likely that prices received or paid by the farmers will be different. It is also likely that characteristics of the farm household population will change, such as the farm size distribution, non-agricultural incomes and household sizes.

*Quantifying Vulnerability.* The AgMIP RIA methods are designed to assess the vulnerability of farm households to possible adverse impacts of climate change. Climate is defined as a probability distribution of weather events that occur at a specific place and during a defined period of time. A change in climate is a change in the probability distribution of weather events. These changes are often described in terms of the mean temperature over a period of time such as a day, month, or year but can also be changes in temperature extremes, the variability of weather events, and other aspects such as rainfall amount and intensity and wind velocity.

Impacts of climate change are quantified as gains and losses in economic well-being (e.g., farm income or per capita income) or other metrics of well-being (e.g., changes in health or environmental quality). In this framework, some or all individuals may gain or lose from a change, and we say, the losers are *vulnerable to loss from climate change*. The AgMIP RIA methodology is designed to quantify the proportion of the population that are losers, as well as the magnitude of loss. It is important to note, however, that in a heterogeneous population, there are typically some gainers and some losers, and thus, the net impact may be positive or negative.

The AgMIP RIA method is designed to quantify climate vulnerability by modeling a heterogeneous population of farm households rather than modeling a 'representative' or average or typical farm. This approach begins with the representation of impacts on the farm household using the concept of economic gains and losses (other metrics of impact can be also be used depending on available data, for example, the impact on health of household members). AgMIP RIA approach uses a statistical representation of the farming system in a heterogeneous region or population to quantify the distribution of gains and losses, for example, due to climate change.

*Quantifying Resilience.* Resilience has been defined in a number of ways in the scientific literature. In ecology, resilience is defined as the capacity of a system to maintain its form and function in response to a shock or disruption. In economic terms, resilience can be defined as the capacity to restore or maintain economic values, such as farm income, or to minimize the loss from an adverse disruption or 'disaster' over the time it takes for a system to return to its 'normal' state. Resilience to climate change can also be defined more broadly as the capacity to cope with change and minimize losses from change and enhance possible benefits of change, and thus can incorporate longer-term responses through adaptation (Antle et al. 2017a).

The definition of resilience as the capacity to withstand disruptions refers to the properties of a given system's performance and is most relevant to analysis of relatively short-term events such as a storm or drought where it can be expected that the system will return to its normal state. In contrast, the capability to adapt or respond by making purposeful changes in a system seems most relevant to longer-term permanent changes in climate and can include adaptations that are designed to improve the capability to withstand shocks or disruptions. Clearly, both concepts of resilience—the ability to minimize the effects of temporary shocks and disruptions as well as the capacity to cope with the long-term shifts in weather patterns associated with climate change—are relevant to analysis of agricultural system performance.

The AgMIP RIA framework can be used to quantify resilience using various sustainability indicators. As noted above, vulnerability is measured as the proportion of farm households that experience a loss over a specified period of time. Loss can be measured in economic terms as reduced income or loss of the capitalized value of income plus assets and also in noneconomic terms such as reduced health or degraded environmental conditions. Similarly, resilience can be quantified by comparing the minimum loss possible with a resilient system to the loss obtained with a less-resilient system. Thus, if a system can achieve the minimum possible loss, its resilience is 100%; otherwise, its resilience is less than

100%. Loss can be measured in economic terms as well as using other environmental or social indicators.

*Representing Future Socio-Economic Conditions.* In a climate change analysis, it is necessary to distinguish between three basic factors affecting the expected value of a production system: the production methods used (i.e., the system technology); the physical environment in which the system is operated, including soils and climate; and the economic and social environment in which the system is operated, that is, the socio-economic setting. In the AgMIP RIA methodology, the non-climate bio-physical conditions and socio-economic conditions are embodied in a Representative Agricultural Pathway or RAP (Valdivia et al. 2015). As discussed further in Chap. 6, RAPs are qualitative storylines that can be translated into model parameters such as farm and household size, prices and costs of production, and policy. The analysis of Questions 3 and 4 defined above is carried out under plausible future conditions defined by Representative Agricultural Pathways. To project the average level of productivity into the future that would occur with ongoing technological advancements (not associated with climate change or adaptation), the AgMIP methodology utilizes the technology trend and price projections developed for global economic models, together with the assessment of technology trends made by research teams in the development of regional RAPs.

## 3.6.2 Farm Household System Models

As Fig. 3.1 illustrates, the farm household system is comprised of bio-physical components (crop and livestock growth, environmental processes such as soil erosion and nutrient transport), economic components of both farm and household decision-making, and social components of the household activities and their consequences.

Crop modelers approach modeling farm production systems as a biological process and typically use experimental data to parameterize and calibrate their models. Economists approach modeling farm production systems from the perspective of the farm decision-maker, typically using observations of actual farm behavior from farm survey or similar data. As we note in Sect. 3.2, the farm and household decision-makers are usually assumed to make purposeful choices to achieve one or more well-defined objectives, subject to the constraints imposed by physical and biological processes and available physical, financial, technological, and human resources. Bio-physical farm system models can be viewed as a process-engineered representation of the production function. Rather than using process-based engineering functions, economists traditionally have used statistical methods to estimate parameters of production functions. Both approaches have strengths and weaknesses (Antle and Stöckle 2017).

An advantage of processes-based models is that they embody relationships that are invariant to a particular site or sample of data. This feature means that they can be used to solve one of the key challenges in ex ante technology impact assessment, that is, to predict 'out of sample' or to predict outside the range of historically observed conditions. Of course, this will only be true to the degree that the models accurately represent fundamental processes, and it is clear from recent research that, there is still a substantial gap between the capability of currently available models and reality. But even when they can reliably predict crop productivity, process-based models have other limitations, notably, they are complex and data intensive, and, as yet, available models do not embody some important elements of production systems such as human labor, the management goals and ability of the decision-maker, and the effects of pests and diseases and pest management. Moreover, models are not available for some important crops and for complex rotations and intercrops and are not well integrated with livestock models.

Strictly empirical, statistical production functions used in many economic studies have the advantage of being able to incorporate the full range of management inputs, and with sufficiently good data, typically, they will do a better job of predicting within-sample for the simple reason that regression-based statistical models provide unbiased estimates of the sample mean. But, these models have many limitations as well. For example, because they do not incorporate the processes that relate soil properties, weather, and nutrients to crop growth, they may provide poor predictions of how yield responds to these factors. Additionally, effects of as-yet-unobserved environmental changes, such as an increase in atmospheric $CO_2$ or ozone, cannot be reliably estimated using observational data. Likewise, they are unable to predict threshold effects outside the range of observed data, such as the effect of temperature extremes or droughts that are not represented in observed data. Available data rarely span sufficiently long time periods to allow reliable statistical estimates of the dynamic properties of production systems.

An example of a detailed production system modeling framework designed for assessing tradeoffs is presented in Box 3.3. This framework combines data from bio-physical simulation models, such as a crop simulation model or an environmental process model, to an economic simulation model. Many models in the literature, often referred to as a 'bio-economic' models, are formulated in this way (see van Wijk et al. 2014 for a survey of such models). If models are available for other household outcomes such as human health or nutrition, the analyst can also simulate these social impacts.

Many farming system models simulate a single, 'representative farm' or a small number of representative types (e.g., a typical 'small' farm and a typical 'large' farm). The problem with this approach is that it does not adequately represent the heterogeneity in the farm population, for example, discussed in Boxes 3.1 and 3.2. Box 3.3 describes a statistical approach that models a population of heterogeneous farm households. For example, in this approach, crop models can be used to simulate the spatial distributions of crop yields caused by varying soil, climate, and farm-specific conditions, and these yield distributions are combined with economic data from the farm population (e.g., farm survey data) to estimate the population mean and variance of economic returns to the production system. 'Scenario analysis' can be used to represent how these distributions would change with an alternative system. With the introduction of a new technology such as a new crop variety, a crop model can be used to simulate the yield distribution among the farms in the population associated with the current variety (System 1 in Fig. 3.8) as well as the new variety (System 2). In the economic analysis, the difference in economic performance between the current System 1 and the prototype System 2 is used to infer the potential adoption rate of the prototype system and the associated economic impacts. Scenarios can also be used to characterize other changes that would affect the adoption of a new technology, such as changes in prices and policy, or changes in the farm population such as an increase in average farm size. Also, environmental and social data can be incorporated in this approach to represent the changes in the distribution of environmental or social outcomes associated with a change in technology.

## Box 3.3 The Tradeoff Analysis Model for Multi-Dimensional Impact Assessment Model

The Tradeoff Analysis Model for Multi-Dimensional Impact Assessment (TOA-MD) is a statistical simulation model designed to carry out ex ante technology assessments of agricultural system sustainability. TOA-MD is based on a parsimonious statistical representation of a heterogeneous population of farm households and is a generic whole farm household model that can integrate farm household data with data from bio-physical farming system models, environmental process models, or field observations from randomized trials or surveys. TOA-MD can be used for several types of analysis: to estimate an adoption rate and impacts of a new technology; to simulate the supply of ecosystem services; and to assess vulnerability to climate and other environmental changes and benefits of adaptive responses to those changes (https://agsci.oregonstate.edu/tradeoffs/applications-library). Impacts are quantified as the 'treatment effects' for those who gain from a change (adopters), those who do not gain (non-adopters), and the full population. TOA-MD is being used in the Agricultural Model Inter-comparison and Improvement project's (AgMIP) regional integrated assessment methodology that combines outputs from global economic models, downscaled climate data, and global and regional socio-economic scenarios with regional and regionally parameterized crop and livestock models to assess climate impacts and adaptations (Antle et al. 2015a, b). (Fig. 3.8)

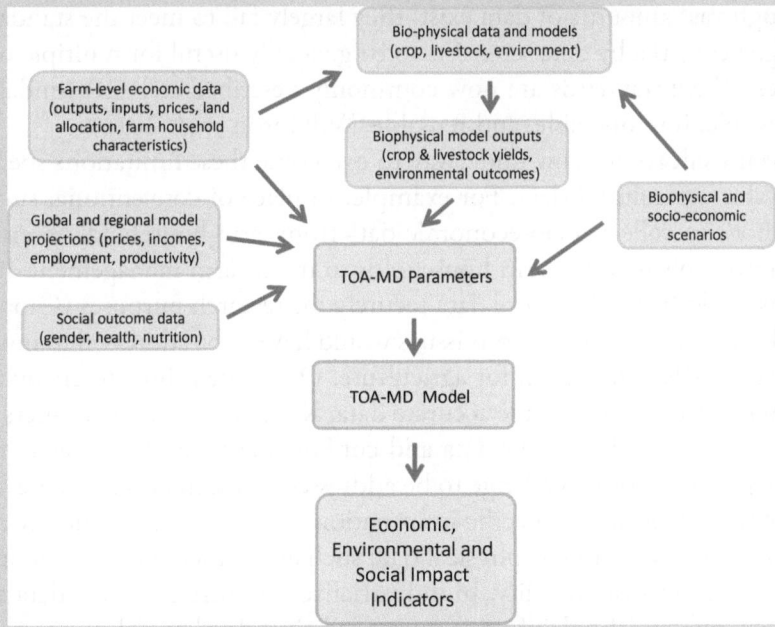

**Fig. 3.8** Tradeoff Analysis Model for Multi-Dimensional impact Assessment (TOA-MD). (Source: Based on Antle et al. (2014, 2015b))

### 3.6.3 Data

The availability of essential data is a critical constraint on agricultural research to develop and test new technologies and to evaluate their productivity and sustainability using models such as the ones discussed in the previous section (Antle et al. 2017b; Capalbo et al. 2017). These data include those generated by experiments, the use of simulation models, data collected directly from farmers using survey methods by researchers and by government agencies, administrative data collected by firms and governments, remotely sensed data from satellites and aerial vehicles, and increasingly, data obtained from sensors located on machinery and even embedded in living plants and animals. Current data systems are designed primarily to serve particular purposes, for example, the agricultural census data collected by many governments are intended to serve political or administrative needs and not to be used for research. Data that are collected by research projects for particular purposes are often not archived or documented so as to be useful for other researchers. As a result, even though vast amounts of data exist, they largely fail to meet the standards judged essential by data scientists to be generally useful for multiple purposes. These standards are now commonly described as FAIR (findable, accessible, interoperable, and reusable; Wilkinson et al. 2016).

Many efforts are now underway to overcome these limitations specifically for agricultural data. For example, in place of conventional survey methods to collect socio-economic data from farm households, decision support tools used by farm households to make farm management decisions could be used to store data securely for research purposes (Capalbo et al. 2017). However, various issues would have to be addressed to implement a FAIR data system for agriculture. One issue is how to encourage farm households to provide accurate data. Key issues for most farmers are the time required to input data and confidentiality of shared data. This issue poses a critical challenge to be addressed if researchers are to be able to obtain the location-specific information needed to link economic and management data to bio-physical data such as weather and soils for analysis of system sustainability. In industrialized countries, private data and related soft and hard infrastructure are being developed by a growing array of management advisory and technology companies, primarily in order to sell advisory services to farms. Data generated by individual producers or by private firms selling data or advisory services are not public and thus not findable or accessible, often even by farmers themselves. There are no established data standards being used, and thus, data are not interoperable even when findable and accessible.

There are also many limitations of currently produced public data. Some of the public data, such as some weather, price, and crop yield data, are freely accessible or available for fees. However, many of the data related to agricultural production are collected for various government administrative purposes and are not intended to be used for research or for private decision-making. For example, much of the data collected by government agencies are findable but are not easily accessible in a timely manner, and then only available in summary or aggregated form, and thus are of limited value for analysis of sustainability issues such as environmental impacts that are site specific. Also, most data fail the interoperability standard, and thus are not usable for artificial intelligence and machine-learning purposes without first being put in a standardized form. A major shortcoming is the lack of detailed, site-specific management and cost of production data. For example, the US agricultural census collects cost of production at the whole farm level but does not collect management or cost information for individual production activities. Another major shortcoming is that none of the available data are collected for the same farms over multiple years (longitudinal data); such data are essential for studying many aspects of system sustainability in economic, environmental, and social dimensions.

Various efforts are underway to address these data challenges. For example, the network of international agricultural research centers (known as the Consultative Group for International Agricultural Research or CGIAR) has launched a Big Data Initiative to develop standards to create FAIR agricultural systems data. The Global Open Data for Agriculture and Nutrition initiative, supported by a number of governmental organizations, is also striving to make governmental data FAIR. In the United States, the National Institute for Food and Agriculture is supporting research on new digital data technologies.

## The Economics of Data and Data Infrastructure

Data are increasingly recognized as a scarce resource, used as inputs into production and decision-making processes, and outputs produced by various types of activities. If data were private goods with well-defined property rights and could be produced and used without externalities, we would expect markets to arise for their efficient production and utilization. However, it is evident that many types of data are not private goods, as their use is non-rival, but may be excludable hence have the attributes of what economists call 'club goods.' Moreover, as yet, property rights are not well-defined for many types of data. Thus, we can say that the 'data market' is substantially in a state of disarray. There are various efforts underway to better define data property rights and related legal and institutional rules and arrangements, but as yet there are many unresolved issues, as evidenced by controversies over the management of personal data by social media firms illustrates. Likewise, for individual farm agricultural data, there are ongoing efforts to define property rights, but as yet no general policies have been established. For example, the US Farm Bureau has promoted a set of 'privacy and security principles for farm data,' and legislation has been proposed to address some of the issues we have identified above related to making government data FAIR.

Similar points can be made about private and public data infrastructure. These issues have already arisen with the development of the infrastructure needed to support universal low-cost access to the internet. Indeed, one of the key limitations to the development of better agricultural data and 'soft infrastructure' is the lack of high-speed internet access by many farmers, even in the technologically advanced regions of the world. This remains a major constraint in the developing world, despite widespread development of cell phone infrastructure. Various developments in sensor technologies create possibilities for on-farm data infrastructure, and one of the key economic questions is the extent to which it will pay farmers to invest in such infrastructure. An important element of a prospective FAIR data system is that on-farm infrastructure could have both private and public benefits if a viable system were in place to enable the public utilization of private data for public good purposes such as research and policy decision-making.

# 3.7  Agricultural Systems as Components of Larger Systems

Chapter 2 discusses agriculture's role in sustainable economic development from an aggregate perspective, and in this chapter, we discuss farm household systems and how we can understand and evaluate their sustainability using tradeoff analysis concepts. Figure 1.1 illustrates that farm household systems are connected to larger ecosystems and socio-economic systems. Understanding these connections remains a major scientific challenge because, as we observed in Sect. 3.1, these interconnected systems are extremely complex, diverse, and heterogeneous. Scientists attempt to meet this challenge by constructing conceptual and computational models, but representing these complex systems mathematically, and obtaining the data needed to construct empirical models that can be used to understand and predict their behavior now and into the future, remains at the frontier of scientific research. Currently, most such efforts involve a modular approach in which physical components (e.g., global and regional climate), biological components (ecosystems or components such as crop and livestock species), and human components (national economies or components such as markets for major agricultural commodities or individual farm households) are connected through key inputs and outputs into each component.

A good illustration of how individual farm household systems are connected to large physical, biological, and socio-economic systems is provided by efforts to assess the impacts of climate change on agricultural systems, as in the AgMIP global and regional integrated assessment approach we discussed above, with agricultural systems at the regional level linked to the global food systems (Fig. 3.7). All farm household systems are impacted by changes in the global climate, albeit differently depending on where they are located. In the case of our commercial wheat farm example (Box 3.1), climate model projections indicate that warmer, wetter winters and hotter, drier summers could increase winter wheat yields by 10–20% but reduce yields of summer crops such as chickpeas by 30% or more. In contrast, in the crop-livestock systems in Machakos, Kenya (Box 3.2), climate projections indicate that maize yields could be reduced by 20–30% in hotter, drier areas; impacts on other crops and livestock have been studied less, but lower and more variable rainfall in semi-arid areas is likely to lower their productivity as well.

How the diverse agricultural systems of the world are connected to regional, national, and global economies also differs greatly. The commercial wheat farms of the US Pacific Northwest are closely connected to global markets through international trade in wheat facilitated by a highly efficient transportation, storage, and processing system. For example, much of the wheat produced in this region is transported by truck to regional storage facilities, transported by rail to ports on the US Pacific coast, shipped by sea to Asian markets, and processed into flour and baked into bread and other foods. Farmers in the US Pacific Northwest are competing in these Asian markets with farmers in Australia and Latin America who are connected to the Asian markets through similar trade networks. Importantly, because these US wheat farms are primarily producing wheat, or in some cases wheat and one or two other crops in rotation, their economic well-being is highly impacted by changes in global wheat prices caused by climate, economic, or policy changes. For example, as discussed in the recent report by the United Nations, commercial wheat farmers in the US and elsewhere are vulnerable to increasingly frequent 'bread basket failures' caused by adverse climate events in multiple major producing regions (IPCC 2019).

In contrast, whereas wheat farmers in the US consume little if any of the wheat they grow, maize farmers in Kenya are likely producing for their own household's consumption and are selling only some—and maybe none—of their maize onto local or national markets. But Kenya exports little if any maize, and thus, its markets are less directly linked to the global maize market which is largely oriented toward livestock feed rather than direct human consumption (Dillon and Barrett 2016). The same is true of the other crop and livestock products in Machakos. Even in the case of maize, Machakos farmers are mainly impacted by local variations in climate, economic conditions, and policy, and those variations have little feedback on markets in other countries because Kenya exports little of the food crops it produces. For example, the major drought in East Africa in 2019 had a large impact on maize production in Kenya and caused local food prices to rise, but this had no global impact; conversely, global 'bread basket failures' have little impact on maize or milk prices in local food markets in Machakos.

## 3.8    Conclusions

In this chapter, we describe farm household systems and their characteristics and the analysis of their performance in the three dimensions of sustainability. We describe the diversity and heterogeneity of agricultural systems in terms of their key characteristics and the implications these features have for the analysis of agricultural system sustainability. Next, we provide an economic rationale for the existence of tradeoffs among economic, environmental, and social dimensions within agricultural systems. We also discuss the limitations of the standard economic approach to evaluating technology and policy options, known as benefit-cost analysis. A key component of tradeoff analysis is the identification of quantifiable sustainability indicators, and we extended the discussion of sustainability indicators and goals, introduced in Chap. 2, to those being used for analysis of farm household systems. The reminder of this chapter focuses on how tradeoff analysis of agricultural system sustainability can be carried out using the regional integrated assessment (RIA) approach developed by AgMIP. RIA is an example of a participatory approach to the development and implementation of sustainable agricultural systems. These processes engage stakeholders to identify goals, sustainability indicators, and technology and policy options, in collaboration with scientific teams who carry out analysis of alternative sustainable development pathways and communicate results to stakeholders, in an ongoing process of innovation and evaluation. We describe the kinds of tools—computer simulation models and data—that are being used by scientific teams to analyze the performance of agricultural system performance along alternative development pathways.

## References

Alkire, S., R. Meinzen-Dick, A. Peterman, A. Quisumbing, G. Seymour, and A. Vaz. 2013. The Women's Empowerment in Agriculture Index. *World Development* 52 (Suppl. C): 71–91. https://doi.org/10.1016/j.worlddev.2013.06.007.

Andersen, E., A.D. Verhoog, B.S. Elbersen, F.E. Godeschalk, and B. Koole. 2006. *A Multidimensional Farming System Typology System for Environmental and Agricultural Modelling*. Linking European Science and Society Report no. 12, May 2006. http://www.seamless-ip.org/Reports/Report_12_PD4.4.2.pdf.

Antle, John, Roshan Adhikari, and Stephanie Price. 2015a. An Income-Based Food Security Indicator for Agricultural Technology Impact Assessment. In *Food Security in an Uncertain World: An International Perspective*. Bingley, UK: Emerald Insight Publishing.

Antle, J., S. Homann-Kee Tui, K. Descheemaeker, P. Masikate, and R. Valdivia. 2017a. Using AgMIP Regional Integrated Assessment Methods to Evaluate Climate Impact, Adaptation, Vulnerability and Resilience in Agricultural Systems. In *Climate Smart Agriculture—Building Resilience to Climate Change*, ed. L. Lipper, N. McCarthy, D. Zilberman, S. Asfaw, and G. Branca. New York: Springer.

Antle, J., J. Jones, and C. Rosenzweig. 2017b. Next Generation Agricultural System Data, Models and Knowledge Products: Synthesis and Strategy. *Agricultural Systems* 155: 179–185.

Antle, J.M., and C.O. Stöckle. 2017. Climate Impacts on Agriculture: Insights from Agronomic-Economic Analysis. *Review of Environmental Economics and Policy* 11 (2): 299–318. https://doi.org/10.1093/reep/rex012.

Antle, J.M., J.J. Stoorvogel, and R.O. Valdivia. 2014. New Parsimonious Simulation Methods and Tools to Assess Future Food and Environmental Security of Farm Populations. *Philosophical Transactions of the Royal Society B* 369: 20120280.

Antle, J.M., R.O. Valdivia, K.J. Boote, S. Janssen, J.W. Jones, C.H. Porter, C. Rosenzweig, A.C. Ruane, and P.J. Thorburn. 2015b. AgMIP's Transdisciplinary Agricultural Systems Approach to Regional Integrated Assessment of Climate Impact, Vulnerability and Adaptation. In *Handbook of Climate Change and Agroecosystems: The Agricultural Model Intercomparison and Improvement Project Integrated Crop and Economic Assessments, Part 1*, ed. C. Rosenzweig and D. Hillel. London: Imperial College Press.

Ballard, Terri J., Anne W. Kepple, and Carlo Cafiero. 2013. The Food Insecurity Experience Scale, 61.

Barrett, C., and M. Carter. 2010. The Power and Pitfalls of Experiments in Development Economics: Some Non-random Reflections. *Applied Economic Perspectives and Policy* 32 (4): 515–548.

Becker, Gary S. 1992. Nobel Price Lecture. Nobel Media AB 2019. https://www.nobelprize.org/prizes/economic-sciences/1992/becker/lecture/.

Capalbo, S.M., J.M. Antle, and C. Seavert. 2017. Next Generation Data Systems and Knowledge Products to Support Agricultural Producers and Science-based Policy Decision Making. *Agricultural Systems* 155: 191–199. https://doi.org/10.1016/j.agsy.2016.10.009.

Coates, Jennifer, Edward A. Frongillo, A. Swindale, Beatrice Lorge Rogers, P. Webb, and P. Bilinsky. 2006. *Advances in Developing Country Food Insecurity Measurement.* American Society of Nutrition. http://pdf.usaid.gov/pdf_docs/pnaeb649.pdf.

Dillon, B.M., and C.B. Barrett. 2016. Global Oil Prices and Local Food Prices: Evidence from East Africa. *American Journal of Agricultural Economics* 98: 154–171. https://doi.org/10.1093/ajae/aav040.

FAO. 2011. The State of the World's Land and Water Resources for Food and Agriculture (SOLAW)—Managing Systems at Risk. Food and Agriculture Organization of the United Nations, Rome and Earthscan, London.

IPCC. 2019. Climate Change and Land: An IPCC Special Report on Climate Change, Desertification, Land Degradation, Sustainable Land Management, Food Security, and Greenhouse Gas Fluxes in Terrestrial Ecosystems. https://www.ipcc.ch/report/srccl/

Jones, James W., John M. Antle, Bruno Basso, Kenneth J. Boote, Richard T. Conant, Ian Foster, H. Charles, et al. 2017. Brief History of Agricultural Systems Modeling. *Agricultural Systems* 155 (July): 240–254. https://doi.org/10.1016/j.agsy.2016.05.014.

Kanter, D.R., M. Musumba, S.L.R. Wood, C. Palm, J. Antle, P. Balvanera, V.H. Dale, et al. 2018. Evaluating Agricultural Trade-offs in the Age of Sustainable Development. *Agricultural Systems* 163: 73–88.

Leff, B., N. Ramankutty, and J.A. Foley. 2004. Geographic Distribution of Major Crops Across the World. *Global Biogeochemical Cycles* 18: GB1009. https://doi.org/10.1029/2003GB002108.

Lowder, S.K., J. Skoet, and T. Raney. 2016. The Number, Size, and Distribution of Farms, Smallholder Farms, and Family Farms Worldwide. *World Development* 87: 16–29.

Nemo, S. 2019. G.E.M.S: An Innovative Agroinformatics Data Discovery and Analysis Platform. https://bigdata.cgiar.org/resource/g-e-m-s-an-innovative-agroinformatics-data-discovery-and-analysis-platform/

Robinson, T.P., P.K. Thornton, G. Franceschini, R.L. Kruska, F. Chiozza, A. Notenbaert, G. Cecchi, et al. 2011. Global Livestock Production Systems. Rome, Food and Agriculture Organization of the United Nations (FAO) and International Livestock Research Institute (ILRI), 152 pp.

Valdivia, R.O., J.M. Antle, C. Rosenzweig, A.C. Ruane, J. Vervoort, M. Ashfaq, I. Hathie, et al. 2015. Representative Agricultural Pathways and Scenarios for Regional Integrated Assessment of Climate Change Impact, Vulnerability and Adaptation. In *Handbook of Climate Change and Agroecosystems: The*

*Agricultural Model Intercomparison and Improvement Project Integrated Crop and Economic Assessments, Part 1*, ed. C. Rosenzweig and D. Hillel. London: Imperial College Press.

van Wijk, M., M. Rufino, D. Enahoro, D. Parsons, S. Silvestri, R. Valdivia, and M. Herrero. 2014. Farm Household Models to Analyse Food Security in a Changing Climate: A Review. *Global Food Security* 3: 77–84.

Wilkinson, M.D., et al. 2016. Comment: The FAIR Guiding Principles for Scientific Data Management and Stewardship. *Scientific Data* 3: 160018. https://doi.org/10.1038/sdata.2016.18.

World Health Organization. 2018. Obesity and Overweight. https://www.who.int/en/news-room/fact-sheets/detail/obesity-and-overweight.

# 4

# Challenges of Sustainable Agriculture in Developing Countries

## 4.1    Introduction

In Chap. 2, we described the growth trajectory of industrialized and developing countries and the role of agriculture in their economic development. Figures 2.3 and 2.5 show that while the contribution of the agricultural sector to GDP is lower in high-income countries, it continues to play a critical role in the transformation of countries from low to high income as suppliers of food, fiber, and fuel, as the source of livelihoods for a large share of the population, as well as labor and other inputs into the initial stages of industrialization. The World Bank uses the Gross National Income[1] (GNI) per capita to classify countries into four broad categories of income—high income (greater than US$12,055), upper-middle income (US$3896–12,055), lower-middle income (US$996–3895), and low income (less than US$996), valued at 2011 prices. This chapter focuses on low- and lower-middle-income economies in sub-Saharan Africa and South and Southeast Asia since close to 90% of the extremely poor population lives in these regions and continues to rely on agriculture.

---

[1] Gross National Income (GNI) includes income generated within a country as gross domestic product (GDP) as well as by income earned by residents living abroad.

© The Author(s) 2020
J. M. Antle, S. Ray, *Sustainable Agricultural Development*, Palgrave Studies in Agricultural Economics and Food Policy, https://doi.org/10.1007/978-3-030-34599-0_4

It also discusses the unique case of China, belonging to the upper-middle-income category yet self-identified as a developing country. This can be understood in view of the fact that, despite its rapid progress in raising average per capita income in recent decades, China's agricultural sector still employs hundreds of millions of low-income people on small farms as in other developing countries. The structure of agriculture in the Latin American and Caribbean regions is diverse in nature with the coexistence of large commercial farms in Argentina, Brazil, and Chile (discussed in Chap. 5) contributing to agricultural exports from the region as well as smallholder farmers. Most countries in the region are heavily urbanized and the share of rural poverty is relatively low compared to countries in sub-Saharan Africa and South and Southeast Asia.

The principles of sustainable development discussed in Chaps. 2 and 3 apply across the World Bank's four income categories. However, the challenges of low-income countries differ from industrialized economies for at least two reasons. First, the urgent human and economic development needs such as eradication of extreme poverty, food and nutrition security, industrialization, and economic growth are unique to the lowest-income countries. Second, the characteristics of the agricultural sectors and systems in developing countries imply distinct pathways to sustainability, even though their design is based on similar principles. In this chapter, we discuss the challenges to move the growth trajectories of agricultural systems and sectors in developing countries in more sustainable directions, and in doing so, hopefully avoid some of the unsustainable pathways that industrialized countries have followed. We describe major agricultural systems and highlight the diversity of agricultural systems in developing countries. We shall see that the low-income and transitional countries are a complex mix of countries at different stages of economic growth. This diversity has far-fetching implications for their agricultural sectors and the design and implementation of pathways toward sustainability.

The major agro-ecological zones of the developing world can be seen in Fig. 3.2. Within these zones, we find a diversity of crop and livestock systems in sub-Saharan Africa (Fig. 4.1), South Asia (Fig. 4.2), and East Asia (Fig. 4.3). Within each farming system is, in turn, a great deal of heterogeneity that cannot be represented in these regional maps. Each of these systems has its own set of sustainability challenges comprising a

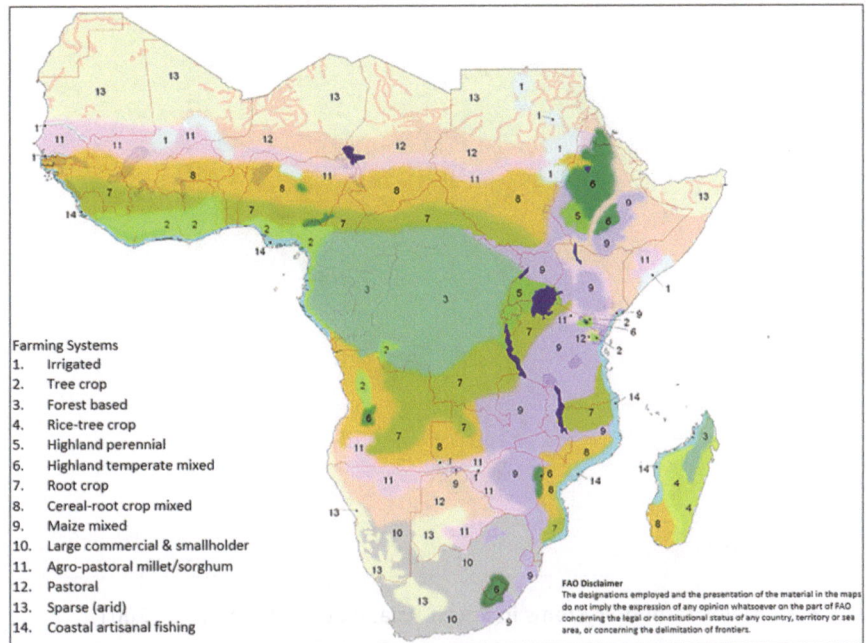

**Farming Systems**
1. Irrigated
2. Tree crop
3. Forest based
4. Rice-tree crop
5. Highland perennial
6. Highland temperate mixed
7. Root crop
8. Cereal-root crop mixed
9. Maize mixed
10. Large commercial & smallholder
11. Agro-pastoral millet/sorghum
12. Pastoral
13. Sparse (arid)
14. Coastal artisanal fishing

FAO Disclaimer
The designations employed and the presentation of the material in the maps do not imply the expression of any opinion whatsoever on the part of FAO concerning the legal or constitutional status of any country, territory or sea area, or concerning the delimitation of frontiers.

**Fig. 4.1**   Major farming systems in sub-Saharan Africa. (Source: Dixon et al. 2001)

subset of the indicators we identified in Chap. 3 (Tables 3.1–3.3). A detailed description of each system and its challenges for development is beyond the scope of this book. We focus on examples that represent some of the most common features. We refer the reader to Dixon et al. (2001) for a more comprehensive description of farming systems across developing regions.

Figure 2.1 shows that most low- and lower-middle-income countries are located in sub-Saharan Africa and South Asia. These regions along with China are projected to be home to 70% of the global population in 2050. The contribution of agriculture in terms of GDP and employment is comparable in sub-Saharan Africa and South Asia and lower in China due to its rapid growth over recent decades (World Bank 2019a). The agricultures of sub-Saharan Africa and Asia can be broadly differentiated on the basis of productivity, intensity in input use, mechanization, and commercialization. We use the examples of low-intensity semi-subsistence

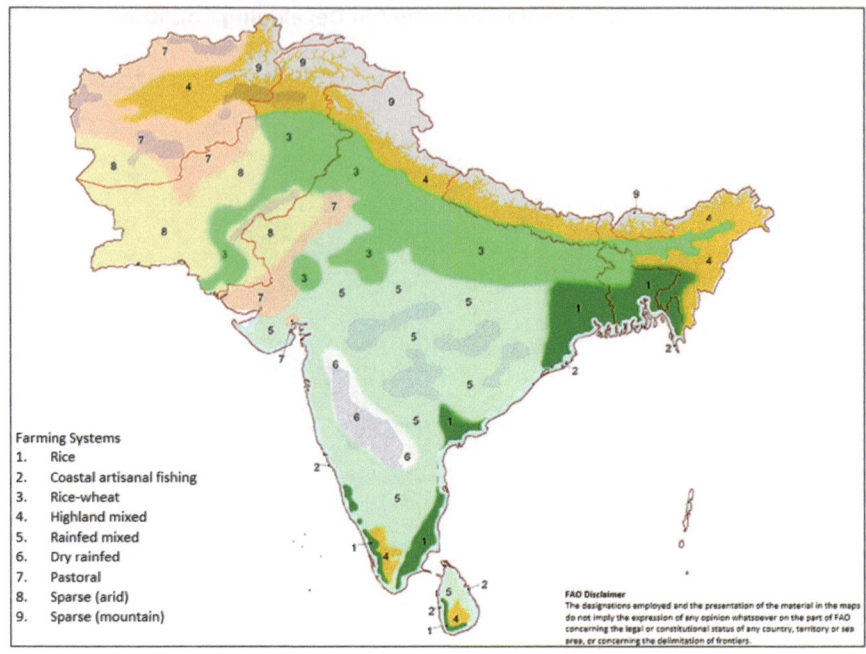

Farming Systems
1. Rice
2. Coastal artisanal fishing
3. Rice-wheat
4. Highland mixed
5. Rainfed mixed
6. Dry rainfed
7. Pastoral
8. Sparse (arid)
9. Sparse (mountain)

**Fig. 4.2** Major farming systems in South Asia. (Source: Dixon et al. 2001)

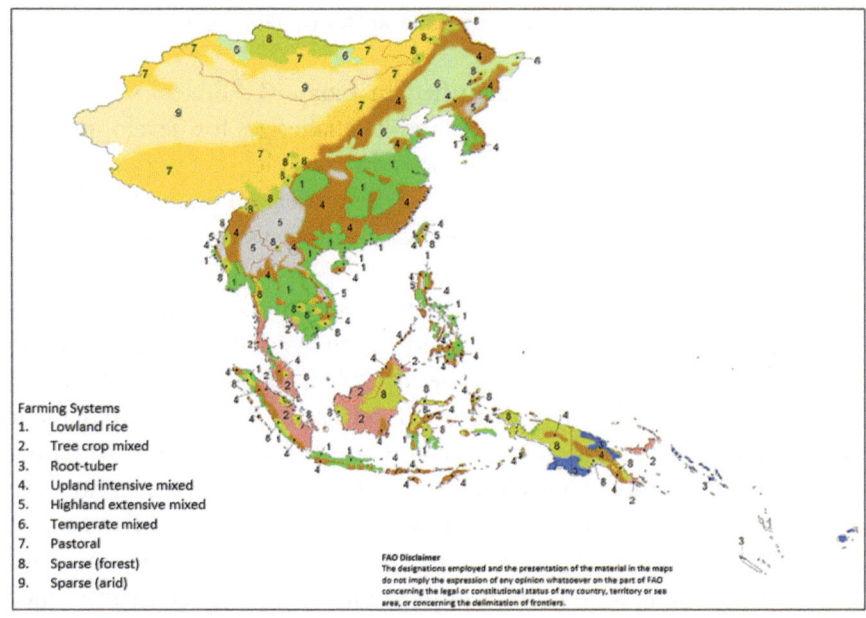

Farming Systems
1. Lowland rice
2. Tree crop mixed
3. Root-tuber
4. Upland intensive mixed
5. Highland extensive mixed
6. Temperate mixed
7. Pastoral
8. Sparse (forest)
9. Sparse (arid)

**Fig. 4.3** Major farming systems in East Asia and the Pacific (Source: Dixon et al. 2001)

systems in Kenya and Tanzania and higher-intensity 'transitional' systems in India and China to distinguish the challenges of subsystems within the developing world. We distinguish transitional systems from semi-subsistence systems since they remain starkly different from industrialized agriculture (Chap. 5), yet have some of the characteristics of more intensive systems. We use these examples to focus on the broad sustainability issues that are distinct between low- and high-intensity production systems within developing countries. It is important to remember that at most levels of spatial aggregation, these examples are an oversimplification. Western Kenya and the Southern Highlands of Tanzania have agricultural systems that resemble high-productivity regions of India. At the same time, subsistence agriculture similar to Kenya and Tanzania is prevalent within most agricultural systems in India. In fact, these 'stylized' examples fall at different points along a continuum of systems with many diverse features and characteristics ranging from the very-low-income, subsistence-level systems to those very similar to the large-scale, commercial systems typical of the industrialized world that we discuss in Chap. 5.

In the next section, we briefly describe the agricultural systems in developing countries. Then, we highlight the current challenges to sustainability in terms of the indicators discussed in Chap. 3. Finally, we describe some of the available strategies for moving agricultural systems in more sustainable directions.

## 4.2 Agricultural Systems in Developing Countries

### 4.2.1 Major Agricultural Systems

Farming systems reflect geographical patterns comprising physical and ecological conditions such as terrain, soil quality, and climate. In this section, we briefly describe the agricultural systems in sub-Saharan Africa, South Asia, and China. The aim is to initiate the reader to the diversity in major farming systems and the within system heterogeneity in these regions. We do not discuss low-income areas of other regions of the world, such as some parts of Western Asia and Latin America, and we refer the reader to the sources below for further information on those other regions. This discussion draws on public resources available from academic literature, publications by the Food and Agricultural Organization (FAO) of the United Nations, the World Bank, and the Consultative Group for International Agricultural Research (CGIAR) (FAO et al. 2019; IFPRI 2014; Hobbs and Osmanzai 2011; Robinson et al. 2011; Dixon et al. 2001).

## Sub-Saharan Africa

The arid and semi-arid regions of the Sahel and Eastern and Southern Africa have low potential for cultivation. These regions are used as rangeland for extensive livestock rearing where animals graze on sparse vegetation. Crop cultivation in these regions is difficult, given the lack of rainfall and unsuitable soils. South of the Sahel, in humid and subhumid regions of sub-Saharan Africa, the majority of small- and medium-scale agricultural households rely on mixed crop-livestock systems. Maize-based cropping systems are common in Eastern and Southern Africa due to its popularity as a staple crop, often intercropped with beans. However, maize is not well suited to nutrient-poor tropical soils unless managed appropriately, because its cultivation rapidly depletes soils without adequate fertilization. Maize is also vulnerable to droughts, pests, and diseases. Other more resilient cereal crops such as sorghum and millet, and root crops such as cassava and sweet potatoes, are also grown throughout sub-Saharan Africa, but are considered inferior to maize by many consumers. Many households also grow banana as a source of starch in their diet and sunflower which is processed as a source of oil.

The type of livestock in the mixed crop-livestock systems has weak regional patterns. In Eastern Africa, cattle density is the highest. Sheep rearing is relatively more common in Western and Southern Africa in arid regions that cannot support cattle for meat or milk. Goat density is highest in Eastern and Western Africa (IFPRI 2014). The mixed crop-livestock system provides a low-cost source of organic fertilizer for cultivation, draft power, as well as valuable assets. In some regions, farmers increase the productivity of their crop-livestock systems using a 'zero-grazing' system that increases their ability to recycle nutrients between crops and livestock (Box 4.5). Animal products provide nutritious food for home consumption as well as high-value marketable products.

Tree- and forest-based cropping systems are found in more humid regions along the equator. Tree-based systems include cocoa, coffee, oil palm, and rubber, which are prevalent in the humid regions of West Africa. Food crops are grown along with trees for home consumption. Incidence of livestock rearing is lesser in these regions. The humid region of Central Africa has vast expanses of rain forests. Forests in the Congo Basin are diminishing due to increasing spread of shifting cultivation with cassava, maize, sorghum, and bean (Dixon et al. 2001).

## South Asia

Broadly, agriculture in South Asia follows a similar pattern as sub-Saharan Africa—livestock-based systems prevail in arid and semi-arid regions without access to irrigation, and mixed crop and livestock systems in regions with higher rainfall and better irrigation facilities. Livestock is closely integrated with most cropping systems in South Asia. Livestock provides cash income, draft power, and manure. Crop by-products are used as fodder for livestock. Livestock systems are diverse; Parthasarathy Rao and Birthal (2008) identified eighteen distinct crop-livestock systems in South Asia. Livestock density in arid and semi-arid regions is lower than in humid areas, but it is more important to livelihoods in drier areas. There is significant variation in the contribution of the type of livestock within agro-ecological zones. Cattle is the most important type of animal across zones and milk the most prominent source of income. Poultry animals are more common in humid regions where feed is more abundant. Sheep are better suited to arid regions while goats are more adaptable to different climatic zones. The incidence of livestock is lower in regions with more irrigation where crops are more productive and profitable.

The Indo-Gangetic Plain (IGP) is a major source of food produced in South Asia. It spreads across semi-arid regions of eastern Pakistan and northwest India to eastern India and most of Bangladesh including southern Nepal. Agriculture in the IGP has high productivity due to heavy seasonal rainfall, rich alluvial soils, and irrigation provided by major rivers in the region. Rice-based cropping systems are prominent in the IGP. In the western IGP, summer rice is grown with winter wheat, often with other crops such as vegetables or coarse cereals. The rice-rice system (rice grown in both seasons) is prevalent in the eastern IGP, and can be integrated with aquaculture in the delta regions of Bangladesh. Northern Bangladesh has greater diversity in its cropping system that rotates wheat, maize, or beans with rice. With abundant rainfall, a third cereal is also cultivated in some areas (Grassini and van Ittersum 2019).

The IGP runs along the foothills of the Himalayas. North of the IGP, highland agriculture is practiced often in terraced systems. In the hilly regions of Nepal located north of India, rice and maize are the two main cereal crops. Rice is grown in terraced systems during the monsoon season. It is sometimes followed with wheat or mustard crop if irrigation is available. Potatoes are also increasingly grown in terrace farming systems. Maize intercropped with finger millet is grown on sloping land. Highland agriculture also includes cultivation of tea, coffee, and spices in eastern and southern India.

In the south of the IGP covering most of central and southern India and Sri Lanka, agriculture is rain fed. These systems grow cereals, oilseeds, and pulses often mixed or intercropped with each other. Legume intercropping helps in the soil's nutrient replenishment by fixing atmospheric nitrogen. Some oilseeds or pulses that need more moisture such as groundnuts, sunflower, pigeon pea, and soybean are grown in the monsoon season. Others such as mustard and safflower are grown as winter crops and coarse cereals and pulses such as millets, sorghum, and chickpea are grown in summer. Cotton-based systems are suitable in regions with black soils in central and south India. It is grown as a cash crop, often intercropped with pulses or cereals (Hobbs and Osmanzai 2011; Dixon et al. 2001).

## East and Southeast Asia

Southeast Asia is predominantly under humid and subhumid agroecological zones. East Asia, mainly comprising of China is a combination of arid desert topography in west and central regions and humid areas in the east. China has about 21% of the global population which relies on 7% of total arable land (Wang et al. 2018). In the arid regions of the Gobi Desert and Qinghar-Tibetan Plateau, less than 1% of the land is cultivated with cotton, wheat, and maize. Agricultural production in China is concentrated in the central and eastern regions, with the Yellow River and Yangtze River providing rich alluvial soil deposits and facilitating irrigation. Agricultural systems in northern China with drier and cooler climate comprising the Northeast China Plain (NECP) and North China Plain (NCP) can be differentiated from Middle-Lower Yangtze Plain (YP), Pearl River Delta (PRD), and Sichuan Basin (SB) which are in eastern and southern China with warmer climates and higher rainfall. Crop intensity increases from northern to southern regions. The NECP typically has a single crop cycle of rice or maize. Double cropping cycles are common in the NCP and SB and also prevalent in the PRD. In the northern regions, rice with maize or wheat is grown due to lesser rainfall and cooler temperature. Cropping systems solely based on rice are popular in southern China. These regions also receive high rainfall while the NECP receives less rainfall and relies on groundwater for irrigation.

Rice cultivation is the primary crop system in several Southeast Asian countries such as Indonesia, Vietnam, Cambodia, Thailand, and Laos (Syuaib 2016). In hilly areas, terraced cultivation of rice is practiced similar to Nepal. The warm and humid climate in the region is suitable for short cropping cycles of rice. Multiple rotations of rice are common with up to three cycles. Upland crop fields grow secondary crops such as vegetables and tubers.

Similar to sub-Saharan Africa and South Asia, mixed crop-livestock systems are prevalent in East and Southeast Asia. Rangeland agriculture in the arid regions of China produces 60% of wool and cashmere and 33% of total milk and mutton production in China by rearing sheep, goats, cattle, and camels. The mixed crop-livestock system is also practiced in areas with intensive agriculture since livestock is a source of nutrients to the soil and draft power for cultivation. Intensive agricultural farms also rear cattle, goats, sheep, and donkeys (Hou 2014). In Southeast Asia, such as in Indonesia and Vietnam, beef cattle is widely integrated into cropping systems.

## 4.2.2  Characteristics

### Farm Size and Employment

Large-scale commercial farms exist in developing countries, particularly in areas with plantation crops such as sugarcane and tree crops, but the majority of the population dependent on agriculture operates 'small' land holdings. The 'small' characterization usually refers to less than 1 hectare, 2 hectares, or sometimes 5 hectares, depending on the productivity of the systems and population density. A study by Lowder et al. (2016) shows that farm sizes in most developing countries have been decreasing consistently since the 1960s. Large sections of the population rely on smaller farms, with fewer people owning larger farms. In Kenya, farmers owning the smallest 25% of land in terms of size have access to only 0.04 hectare per person (Djurfeldt and Jirström 2013). In the 1990s, average farm size in sub-Saharan Africa was 2.4 hectares which declined to 2.16 hectares in 2008. Nigeria is the only country in sub-Saharan Africa where farm sizes have increased. In South Asia, there is an increasing concentration of farms less than 2 hectares due to land fragmentation (Lowder et al. 2016). The average farm size in India has been reported to be slightly more than 1 hectare and decreasing (Chand et al. 2011). While the declining trend in farm size has been observed across several Indian states, Fig. 4.4 shows that in Punjab, where agricultural productivity is high, there has been an increase in the share of larger farms (> 4 hectares) since 1990. Haryana, which is also a highly productive state, has recorded a decline in the share of large farms. Figure 4.4 also shows the prevalence of small farms (< 2 hectares) across several states except Punjab and Rajasthan. Rajasthan is located in arid and semi-arid regions where productivity is low and extensive agriculture is practiced with larger farms (ICRISAT 2015). Farm sizes in arid and semi-arid areas tend to be larger than in humid areas. Generally around the world, we see that agro-ecological zones with higher rainfall have higher agricultural productivity and thus we see a smaller minimum farm size that can provide a subsistence income, compared to arid regions where the productivity is lower and minimum size is typically larger.

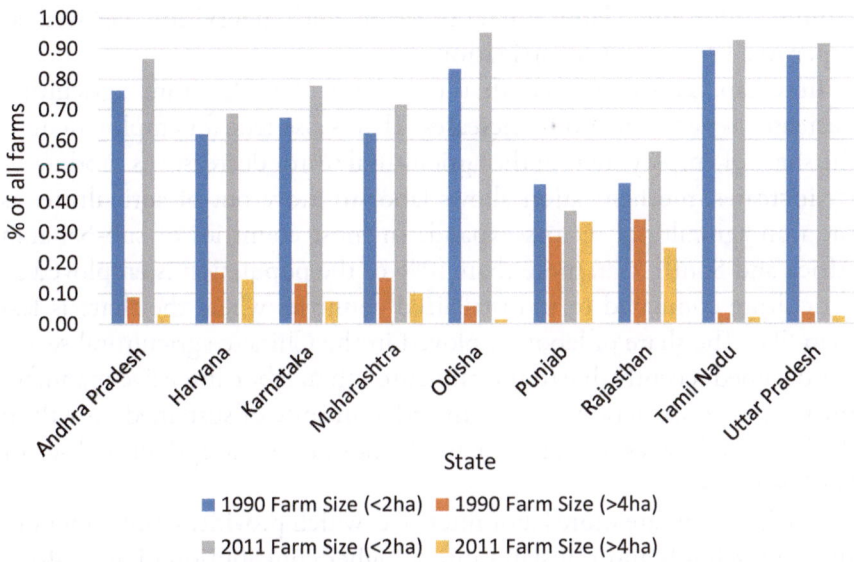

**Fig. 4.4** Change in farm size in key Indian states. (Source: Author's calculation based on ICRISAT 2015)

The average farm size in China was reported to be 0.6 hectare in 1997 and continued to decline into the 2000s (Chand et al. 2011). In spite of the small farm sizes, agricultural productivity in China has been higher than other Asian countries. In recent years, sustained growth in the non-agricultural sector has led to increased migration from rural to urban areas and increase in farm size, although the increase has been slow with 70% of farms still smaller than 2 hectares (Djurfeldt and Jirström 2013). Southeast Asian countries also have a high concentration of small farms. In the early 2000s, almost all farms in Vietnam and about 85% farms in Indonesia were less than 2 hectares in size. Farms more than 5 hectares in size are about 10–20% of farms in Myanmar, Philippines, and Thailand (OECD-FAO 2017).

Landholdings by agricultural households are often fragmented over generations. A farmer owning 1 hectare of land (roughly equivalent to one football field) may do so in several dispersed parcels. FAO reported that in 2005, the average number of parcels in Africa and Asia were 3 and 3.2, respectively, while in North and South America, it was 1.2 (FAO 2010). Fragmented landholding makes mechanization difficult, requires additional labor for cultivation and supervision, as well as expensive investment for irrigation.

In Chap. 2, we showed that the share of GDP from agriculture decreases as per capita GDP increases. That is also true for employment—the share of employment in the agricultural sector decreases as productivity increases, mechanization allows labor to move out of agriculture, as the non-agricultural sectors expand. In most countries of sub-Saharan Africa and South Asia, more than 40% of the population is employed in agriculture compared to industrialized countries where the share is less than 5%. The share of labor employed in the Chinese agricultural sector has declined recently due to the rapid growth in labor-intensive manufacturing and construction sectors. In India, in spite of sustained growth in the last two decades, the movement of labor out of the agricultural sector has been slow.

Smaller farms are more labor intensive, which provides employment to rural households and translates into a higher contribution of agriculture to a country's GDP and employment. With increasing advancement of the non-agricultural sector, gainful employment in sectors such as manufacturing, transport, and construction becomes available, attracting agricultural labor into non-agricultural sectors. This out-migration can lead to consolidation of farm land, higher agricultural wages, and greater investment in mechanized agriculture, and higher productivity. In semi-subsistence systems, this transformation is in its nascent stages, whereas this process is ongoing in transitional economies.

## Productivity and Input Use

Agricultural productivity in South Asia has benefitted from the Green Revolution (GR) that began there in the 1960s. The GR focused on the adoption of high-yielding rice and wheat varieties, along with chemical fertilizers, pesticides, herbicides, and irrigation. The average cereal yield in South Asia (about 3200 kilogram/hectare) was twice that of sub-Saharan Africa (less than 1500 kilogram/hectare) in 2017. The highest yields are recorded in the Indian state of Punjab, in the northwest (NW) IGP, exceeding 4500 kilogram/hectare for wheat and around 4000 kilogram/hectare for rice and maize in 2015–2016. In the highly populated states of Bihar and Uttar Pradesh, also belonging to the rice-wheat crop-ping system in the IGP, rice and wheat yields are 40–50% lower than in Punjab. Productivity of other food crops such as pulses and coarse cereals also lags behind in the rest of the country (MOSPI 2019).

In comparison, the average rice and maize yields in China exceeded 6000 kilogram/hectare by 2013, lower than the United States but com-parable to Western Europe (Wu et al. 2018). There has been a spatial shift in regional grain productivity in the 1980s which continues in the 2000s. The NECP and NCP regions north of the Yellow River and parts of west-ern China along the banks of the Yellow River have experienced sharp increases in cultivated land and grain output. Wang et al. (2018) show that between 2003 and 2014, all counties doubling output growth were concentrated north of the Yellow River. While the rice-based cropping systems prevail in the traditional agricultural regions in southeastern China, the regions expanding agricultural production have focused on maize cultivation in the semi-arid regions of north and western China.

Input use differences between China, South Asia, and sub-Saharan Africa follow a pattern similar to yield. Fertilizer consumption rates in South Asia (160 kilogram/hectare) were ten times higher than the average in sub-Saharan Africa (16 kilogram/hectare) in 2016. In China, fertilizer and pesticide use is the highest in the world. It uses 30% of world's fertil-izer on 9% of global arable land. China's fertilizer use rate exceeds 500 kilogram/hectare and pesticide use exceeds 16 kilogram/hectare. Empirical evidence has shown that small farms tend to use more fertilizer and pesticides (Wu et al. 2018). Fertilization rates vary substantially within regions and for crops. In China, between 1998 and 2008, fertil-izer consumption for horticulture crops increased rapidly and contrib-uted to 30% of overall fertilizer use in 2008 compared to 18% in 1998. The share of fertilizer demand for grain production fell from 71% to 58%. The Huang-Huai-Hai Plain and Henan province typically recorded the highest fertilization rates for grains. During the ten- year period, as grain production expanded into the northern regions of China, grain fertilizer consumption in the Northeast Plains increased substantially. Western China also showed high growth rates in grain fertilization (Xin et al. 2012). In India, higher fertilization rates are observed for both food crops such as rice and wheat in the IGP as well as nonfood crops such as cotton. Regionally, higher fertilization is observed in more productive areas of northwest India, particularly states of Punjab and Haryana compared to the rest of the country (FAO 2005). Chemicals such as pesticides and herbicides are widely used to tackle pests and weeds.

Agricultural systems in sub-Saharan Africa use significantly less purchased inputs compared to South Asia and China. In 2016, commercial fertilizer consumption was 16 kilogram/hectare on average (World Bank 2019c). However, regional averages disguise heterogeneity in application rates in specific regions. Analysis of six African countries shows substantial variation in agro-chemical application rates in the region. Situations of relatively high and even 'over-use' (meaning, uneconomic) are observed in the higher potential zones of Kenya, Malawi, Ethiopia, and Nigeria. In less productive regions, input use is constrained by high prices and transport costs, lack of availability when needed, unreliable weather, and lack of experience. Adoption of improved or hybrid seeds is widely prevalent in South Asia but their adoption rate in sub-Saharan Africa is low in many areas, for example, ranging between 9% and 22% in Ethiopia and Niger and less than 30 % in the lowland areas of Kenya (Sheahan and Barrett 2017). In the semi-arid low-productive system in Machakos, Kenya (Box 3.2), input use is lower for staple grain crops grown for home consumption compared to vegetables which are grown as cash crops. The trend is similar for other inputs—pesticides and irrigation—where more investments are made for vegetables compared to maize. Less than 20% of maize production is irrigated, thus the majority relies on erratic rainfall in semi-arid conditions (Valdivia et al. 2017).

Figures 4.5 and 4.6 show the variation in maize yield from hybrid and traditional seeds and fertilizer use intensity in Kenya and Tanzania between high- and low-productive agro-ecological zones. Both figures show that yield and intensity of fertilizer use are persistently low in unfavorable agro-ecological zones. Yield increases slowly as crop varieties are improved and fertilizer use is increased. Higher productivity is achieved with a combination of hybrid seeds and higher fertilizer use in regions with favorable agro-ecological conditions. This combination led to maize yield of 2500–3000 kilogram/hectare between 2007 and 2010 in Kenya and 1200–1700 kilogram/hectare during 2008–2014 in the more productive regions of Tanzania.

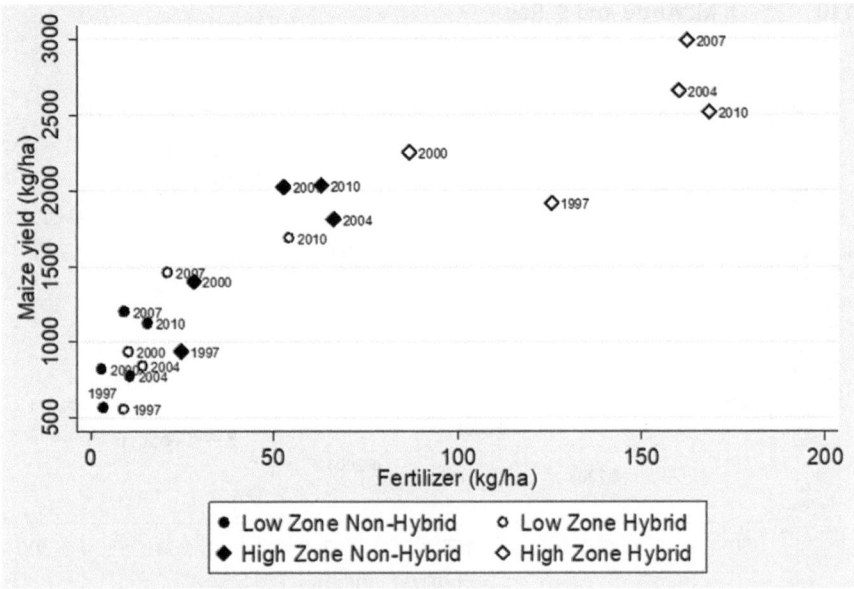

**Fig. 4.5** Maize yield and fertilizer use in Kenya. (Source: Authors' calculation based on Tegemeo Institute 2019)

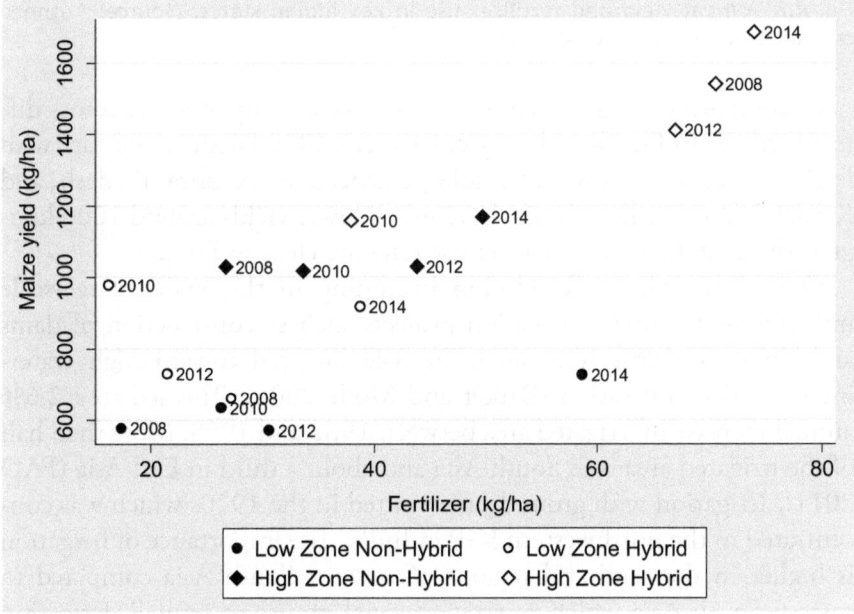

**Fig. 4.6** Maize yield and fertilizer use in Tanzania (Source: Authors' calculation based on Living Standards Measurement Survey, Tanzania (NBS 2011–2015))

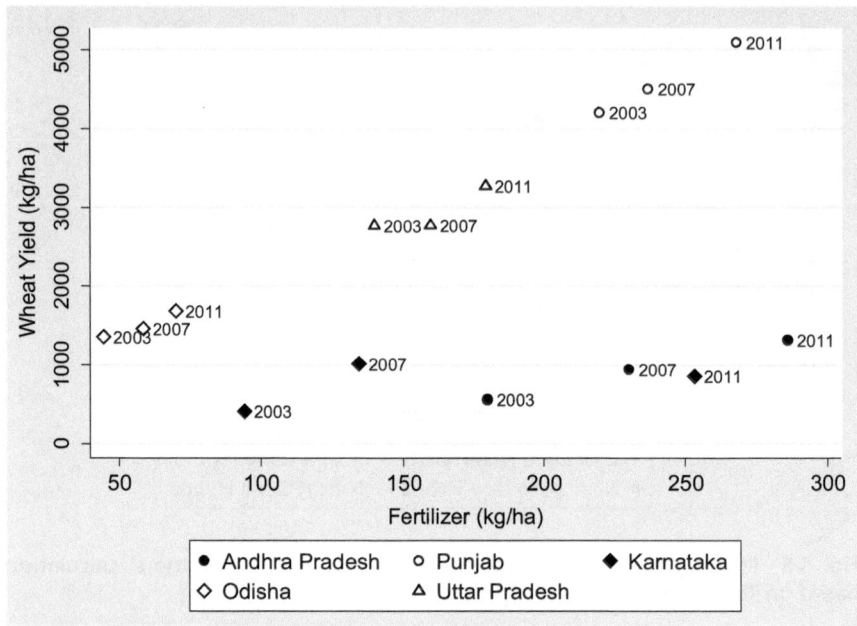

**Fig. 4.7** Wheat yield and fertilizer use in key Indian states. (Source: Authors' calculation based on ICRISAT 2015)

In comparison, Fig. 4.7 shows wheat productivity in India across different states. In Punjab, wheat yield exceeds 4000 kilogram/hectare with fertilizer use of around 250 kilogram/hectare. Andhra Pradesh and Karnataka, in southern India, have much lower yields around 1000 kilogram/hectare, however, fertilizer use rates are close to Punjab.

During the Green Revolution beginning in the 1960s, large-scale investments in surface irrigation projects such as construction of dams and canal networks were made in Asia targeted toward high water-intensive rice cultivation (Barker and Molle 2004). This led to a 2.6% annual increase in irrigated area between 1962 and 1998. More than half of the irrigated area is in South Asia and about a third in East Asia (FAO 2011). Irrigation with groundwater started in the 1970s which was concentrated in the northwestern IGP in India. The importance of irrigation is higher in the semi-arid regions of western South Asia compared to eastern South Asia which receives seasonal monsoon rainfall. Irrigation

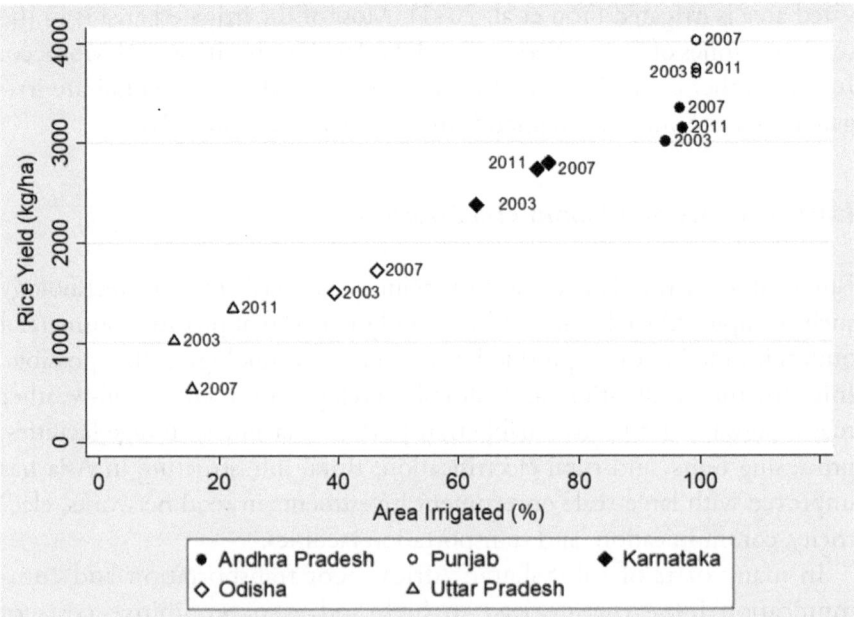

**Fig. 4.8** Rice yield and percent area irrigated in key Indian states. (Source: Authors' calculation based on ICRISAT 2015)

demand is also higher in the drier months of November to March when wheat is grown. Indian states with better access to irrigation are able to achieve higher productivity. Figure 4.8 shows that Punjab, with close to universal access to irrigation, produces around 4000 kilogram/hectare, compared to Odisha and Madhya Pradesh where rice yield is less than 2000 kilogram/hectare and less than half of the area was irrigated.

Chinese agriculture is also heavily reliant on ground water irrigation. Southeastern China receives the heaviest rainfall, which decreases toward the north. The NCP which produces almost half of the food grains in the country relies on groundwater irrigation with cheap electrical pumps. Since most of the rainfall is received in the summer months, the relatively high productivity of wheat in the drier months relies on groundwater irrigation (Zhang 2011).

Sub-Saharan Africa is well endowed with rich natural resources with vast areas of underutilized agricultural land. Only about 4% of the total cultivated area is irrigated (You et al. 2011). Most of the irrigated area is in the semi-arid zones of Egypt, Sudan, South Sudan, South Africa, and Morocco. In some other countries, pockets with high agricultural potential are irrigated such as Southern Highlands in Tanzania and western Kenya.

## Infrastructure and Commercialization

Rural infrastructure is necessary to facilitate farmers' access to technology such as improved seeds and fertilizers, and markets that provide commercial outlets for the increased production that better technology makes possible. Infrastructure facilitating agricultural development includes all-weather roads, railways and ports, reliable transportation facilities, storage facilities, processing units, and rural electrification. Rural infrastructure in Asia has improved with large-scale government investments in road networks, electricity, communication, and transportation facilities.

In many parts of sub-Saharan Africa, poor transportation and communication infrastructure lead to high and even prohibitive costs of accessing markets, both for the purchase of agricultural inputs and for marketing agricultural products (Pingali et al. 2014). Without access to markets, processing facilities, or good storage facilities, farmers lack assurance of returns on their investments in technology such as improved seeds, fertilizers, animal breeds, and feed. This poses particularly severe challenges in the commercialization of perishable animal products such as eggs and milk as well as fresh fruits and vegetables. One fact that limits economic development in sub-Saharan Africa is that it has the largest number of landlocked countries in the world. In addition, restrictions on intercountry trade often limit the opportunity to access markets in neighboring countries. Lack of infrastructure also isolates markets and limits the movement of food into areas of scarcity, often exacerbating local food shortages caused by droughts and other disruptions. Several African nations, under the Comprehensive Africa Agriculture Development Program (CAADP), are committed to invest in infrastructure development that will increase mobility, facilitate inter- and intracountry integration, as well as provide wider markets for their agricultural produce.

The lack of rural infrastructure can partly explain the subsistence nature of smallholder agriculture in much of sub-Saharan Africa and other areas of the world where subsistence agriculture persists. Investment in purchased inputs such as hybrid seeds, fertilizers, and pesticides can be difficult to recover if the final commodity does not have a market. Indeed, without access to markets for increased production, adoption of more productive technologies is likely to drive down prices in isolated markets, thus offsetting much of the gains from adoption. In addition, where weather risk is high, risk-averse farmers see a reduced incentive to invest in purchased inputs that may perform well under good conditions but may fail to perform any better, or even worse, than traditional crops and livestock.

## 4.3  Current Status Using Sustainability Indicators

### 4.3.1  Economic

In low-income populations, per capita income is a key indicator of economic well-being. Income, meaning households' 'real' net income from farm and non-farm sources, can be challenging to measure for agricultural households who consume some of their own production and have family members who work on the farm, because the value of auto-consumed food and self-provisioned labor is unobservable. The economic status of a household can also be described in terms of its wealth, often proxied with privately owned productive or financial assets such as farm equipment and livestock.

Of particular concern are the extremely poor households, typically indicated by the head count poverty rate defined as the percentage of farm households with per capita incomes below a poverty line. Most countries have their own poverty lines, based on local living standards and values. These national poverty lines vary widely across countries. To try to make meaningful comparisons across countries, international organizations such as the World Bank that collect and report such information utilize poverty lines based on estimates of the 'real' income needed to meet basic human needs (food, shelter, clothing, etc.), adjusted for differences in cost of living across countries.

For example, the international poverty line (IPL) developed by the World Bank and used by many organizations is US$1.9 per day per person (in 2011 international, i.e., purchasing power parity-adjusted, terms). Figure 4.9 shows the dramatic fall in the absolute number of the extremely poor from 36% of the world population in 1990 to less than 10%in 2015. Countries in East Asia and the Pacific have made remarkable improvements in poverty reduction. In 2015, 41% of the people in sub-Saharan Africa and 16% of the people in South Asia were extremely poor. Current trends suggest that by 2030, nine out of ten extremely poor people will live in sub-Saharan Africa.

Crop or livestock productivity is another key indicator closely related to income and other important indicators such as food security. At the farm household level, crop and livestock productivity are typically measured in terms of partial productivity measures such as crop yield (i.e.,

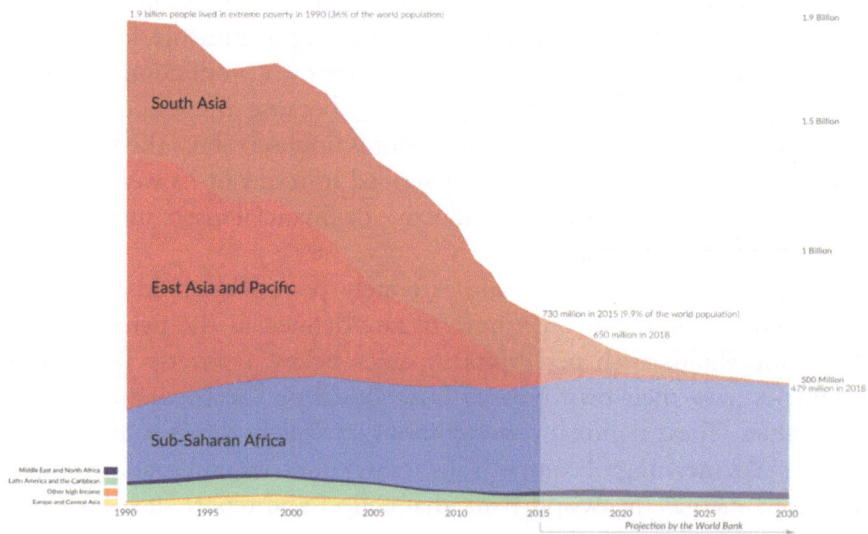

**Fig. 4.9** Number of people in extreme poverty until 2015 and projection for 2030. (Source: Roser and Ortiz-Ospina 2013)

the quantity produced per land area), or for livestock, production per animal or per animal unit (often a standardized 'tropical livestock unit' is used that translates all types of livestock into per-cow equivalent units). Another measure is the yield gap, defined as the difference between potential yield, usually estimated from experiments with 'optimal' agronomic management, and the realized yield on a farmer's field. These gaps in developing countries have been large and persistent. The Global Yield Gap Atlas (GYGA) shows maps of major food crops in different countries for specific agro-ecological zones to show the potential to increase food production across developing economies. It shows that across South Asia, yield gap in wheat production exceeds 50%. The gap is higher for rain-fed wheat cultivation. In sub-Saharan Africa, relative yield gap for rain-fed maize exceeds 80% in several countries. For other coarse cereals, legumes, and vegetables, the yield gaps are higher. In China, given the relatively higher yields in comparison to other developing regions, the yield gaps are less wide. The yield gap in irrigated rice production in southern and North China Plains ranges up to 40%. Yield gaps for maize production are larger, with a greater share of the region in the 40–50% range. Maize production in northern and northwestern China records very low yields relative to its potential (Grassini and van Ittersum 2019).

The potential yield of the rice-wheat system in the productive regions of the IGP such as Punjab was estimated to be 10.6 tons per hectare for rice and 7.7 tons per hectare for wheat (Aggarwal et al. 2000). Since the late 1990s, yield in the region has stagnated. In 2015–2016, the yield gap remained around 40–60% with rice yield less than 4 tons per hectare and 4.5 tons per hectare for wheat (MOSPI 2019). In China, yield has also stagnated since the late 1990s (Xin et al. 2012). Climate change poses further challenges to sustain agricultural yields by increasing the likelihood of sudden weather shocks (Box 4.1).

Different reasons explain the relatively low productivity of the agricultural systems in developing countries. For instance, the Machakos region in Kenya has a semi-subsistence system where households' consumption greatly relies on their farm production for the simple reason that their incomes derive mainly from farm production. Farmers prefer to grow local varieties for home consumption, and farmers can save their own seed for subsequent crops or buy them at low cost in local markets. These local varieties have low yields but may be less vulnerable to droughts than more expensive higher-yielding hybrid varieties, particularly in arid conditions without access to irrigation. In these systems, cereals are often intercropped with other legumes and vegetables for home consumption, with legumes providing some nitrogen to other crops due to their ability to fix atmospheric nitrogen in the soil.

### Box 4.1   Impact of Climate Change on Agriculture

Agriculture, by its nature, is heavily dependent on bio-physical, ecological, and weather phenomena. Thus, it is directly affected by climate change. The impact of climate change in agriculture is being observed across the world in terms of shifting weather patterns, change in habitat for animals and marine life, and more frequent sudden adverse weather events which can directly (as in the case of droughts and floods) or indirectly (due to new pests and diseases) lead to crop loss and death of livestock. In developing regions, the physical impact of climate change is likely to be higher, since they are more reliant on the agricultural sector and have lesser economic resources and coping strategies (Mertz et al. 2009).

Predicting the impact of climate change on agriculture is complex since it involves the interaction of the climate, economic, and social systems over long periods of time. Scientists have developed computer simulation models to implement this type of analysis using 'simulation experiments.' These simulations are driven by projections of greenhouse gas emissions called 'representative concentration pathways' (RCPs) and projections of socio-economic conditions called 'shared socio-economic pathways' (SSPs).

Nelson et al. (2014) report simulations from ten different modeling groups around the world that use computer simulations of the global food system. They all use climate, crop, livestock, and economic models along with a range of RCPs and SSPs to project possible impacts on agriculture. These 10 models generate a wide range of different possible outcomes, but predict, on average, an 11% reduction in agricultural yields and a 3% reduction in consumption globally by mid-century. Regional results show that the largest yield losses are likely to be in sub-Saharan Africa, close to 20% on average. Potential reductions in consumption are the highest in India, followed by sub-Saharan Africa and Southeast Asia. Globally, a 20% increase in major commodity prices is predicted, implying that the poorest people, who disproportionately live in developing regions and spend a large share of their income on basic foods, will be the most adversely affected by climate change impacts on agriculture.

In the more productive areas in sub-Saharan Africa, such as western Kenya and Southern Highlands of Tanzania, productivity is higher due to favorable soil and weather conditions. In transitional systems in northwestern IGP in South Asia and eastern regions of China, after the dramatic gains in productivity during the GR, productivity has stagnated due to the highly input-intensive production practices. Repeated nutrient-intensive cropping cycles have depleted soil nutrients, high rates of macronutrient fertilization have led to micronutrient deficiencies, nutrient leaching into the water table has increased levels of soil salinity, and over-exploitation of groundwater has severely depleted the groundwater tables. Thus, the agricultural system is unable to support further productivity increases and has shown signs of decline in productivity. Furthermore, other regions in India and China are modeling the high-productivity regions leading to the spread of these unsustainable agricultural practices.

Water demand in Asia is projected to increase by 30–40% in 2050 compared to 2010 (Satoh et al. 2017). Current areas under water stress such as north India, Pakistan, and north China will severely worsen. Rice cultivation in South Asia and China is highly water intensive. In northwestern IGP, the rice-wheat cropping system consumes about 11,650 cubic meter water per hectare. One kilogram of rice is produced with 5000 liters of water (Bhatt et al. 2016).

## 4.3.2 Environmental

Soil health and water quality across the developing world are deteriorating as a result of unsustainable farm management practices. For example, the highly water-intensive farming system in the western IGP has led to a decline in the regional groundwater table at an alarming rate. One estimate during 2008–2012 reported that the water table in Punjab fell by 70 centimeters (more than two feet) each year (Gulati et al. 2017). As a result of the falling water tables, wells and pumps are drilled deeper into the aquifer and powered with cheap electricity. In Punjab between 1998 and 2012, carbon emissions increased by 110% in some areas (Dhillon et al. 2019). Similar water stress is prevalent in the semi-arid regions of northern China where water availability per person has fallen and ground water use has increased.

The high rates of fertilization in IGP are often inefficient, with poor timing, wasteful methods of application, and applications made without soil tests to inform the necessary nutrient requirements. Fertilizers address macronutrient deficiencies of nitrogen, phosphorous, and potassium (NPK) leading to micronutrient deficiencies in iron, zinc, and manganese. These practices have led to declines in soil quality in the region and yield stagnation or decline. Alternating between puddled rice cultivation (tillage in flooded fields) and intensive tillage to prepare well-drained soil for wheat cultivation leads to the formation of a dense layer of subsoil (hardpan) that hinders root formation for the wheat crop and limits nutrient and moisture absorption at deeper soil layers (Chauhan et al. 2012). The crop system's response to fertilizer has reduced over time while farmers have continued to increase application rates. For example, in Punjab, nitrogen use efficiency for rice production is only 30–40%. Nutrients not taken up by crops are lost due to surface runoff, or leach into groundwater and contaminate drinking water sources in Punjab as well as neighboring states, causing severe health impacts (Bhatt et al. 2016; Chauhan et al. 2012; Pingali and Shah 2008). Arsenic contamination of groundwater, associated with cancer, heart diseases, and birth defects in children, has been reported across the IGP including north India, southern lowlands of Nepal, and Bangladesh (Virk 2018; Mueller 2017; Harvey et al. 2002).

The Green Revolution (GR) in Asia led to continuous cropping of rice over long periods of time without fallow cycles. The focus of the Green Revolution in the IGP was to increase food production to avert looming food scarcities. However, the widespread adoption of the rice-wheat cropping system drove out other crops including sorghum and millet, as well as increased use of chemical pesticides. In Nepal, the GR has led to deforestation to increase agricultural land under mono-cropping. It has also replaced indigenous farming practices that preserved biological diversity in the systems with yield-intensifying technologies (Upreti and Upreti 2002). Preservation of biodiversity has an intrinsic value, and can improve the resilience of farming systems by maintaining natural predators to pests.

Soil quality across sub-Saharan Africa has deteriorated over time due to adverse agroclimatic conditions and repeated cultivation of nutrient-demanding crops without adequate soil replenishment either through traditional long fallow periods or application of sufficient quantities of organic or chemical fertilizers. Figure 4.10 shows that regions in East Africa and Central Africa suffer from significant soil nutrient constraints. Aluminum toxicity due to continuous leaching of nutrients is one of the widespread problems across the continent.

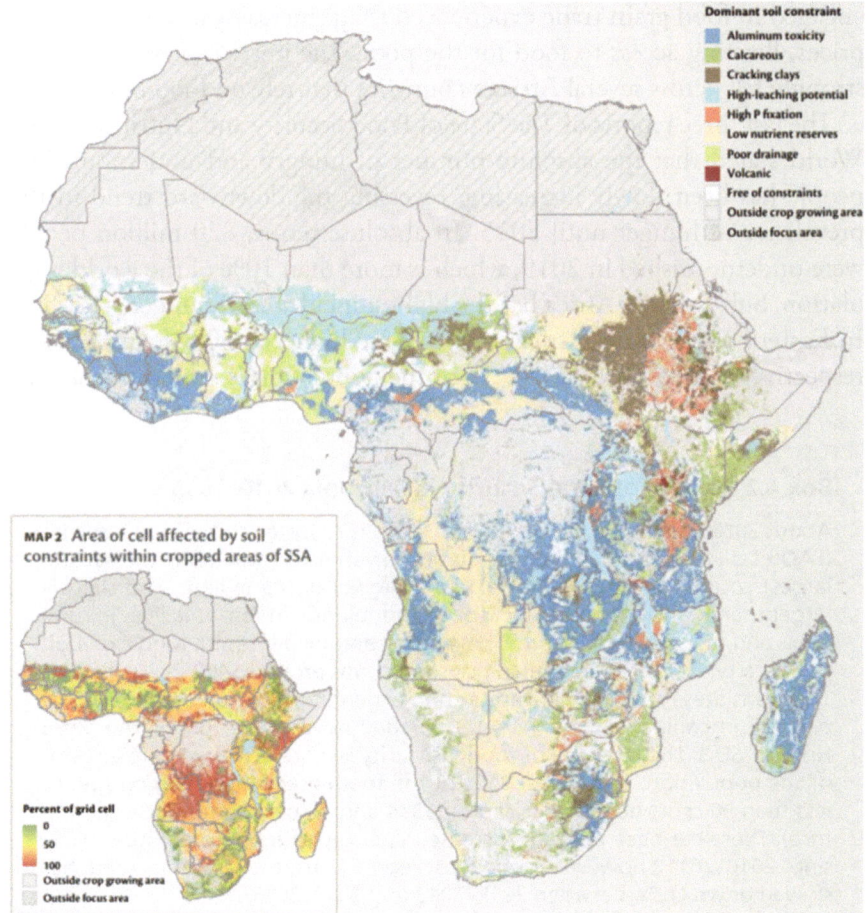

**Fig. 4.10** Soil constraints map of Africa. (Source: IFPRI 2014)

## 4.3.3  Social

Several indicators are used to measure food security to capture its multidimensional nature. At the macro level, physical quantities of food are used to evaluate the food self-sufficiency of a country. Often food acquired via trade is also included in the measure of total food availability. While trade can smooth production shocks faced by a country, overreliance on trade can make a country's basic domestic food supply vulnerable to international price shocks. During the 2008–2012 food price crisis, countries engaged in food grain trade experienced sharp increases in domestic food prices, limiting access to food for the poor. The impact of this crisis was strongly felt across several African countries that rely on food imports.

The last three reports of The State of Food Security and Nutrition in the World stated that the absolute number of hungry and undernourished people has been slowly increasing, reversing the downward trend in the prevalence of hunger until 2015. In absolute terms, 820 million people were undernourished in 2019, which is more than 10% of the world population. Sub-Saharan Africa has the highest incidence of hunger at 22.8%. In Eastern and Central Africa, the rates are higher at 30.8% and 26.4%, respectively. In South Asia, 14.7% of the population is undernourished.

---

**Box 4.2   Status of Food Security in Tanzania 2010–2015**

About 30% of the population in East Africa is reported to be food insecure (FAO et al. 2019). Tanzania's population of about 57 million is the second largest country in the region next to Ethiopia at 105 million, and slightly larger than Kenya. A total of 66% of the population in Tanzania lives in rural areas and relies on agriculture. The median annual per capita food expenditure in a typical rural Tanzanian household was around $460 in 2014–2015. This translates to a daily per capita food expenditure of approximately $1.3 per day in comparison to the $1.9 per day poverty line under the 'Zero Hunger' SDG. The intensity of food insecurity is more acute in some sections of the population. The lowest 25% of the food-insecure population spends less than 90 cents per day per capita ($314 annual per capita food expenditure). There has been a steady increase in per capita food expenditure (50% since 2010–2011); however, in the most recent years, the rate of increase has slowed down (12% between 2012–2013 and 2014–2015).

*(continued)*

**Box 4.2 (continued)**

Food security measured in terms of the Food Insecurity Experience Scale (FIES) shows that 31% of the population is moderately or severely food insecure, of which 14% is severely food insecure (Table 4.1). Moderately food-insecure households switch to poorer-quality foods, leading to lower nutritional intake, and may face psychological stress due to their vulnerability to further food insecurity. Severely food-insecure households also reduce their food consumption by skipping meals and reducing their meal sizes, or go without any food consumption in a day. The reduction in diet quality is reflected in the Household Dietary Diversity Score (HDDS). On average HDDS for severely food-insecure households shows that they consume one less food group compared to the food-secure households. At a more disaggregated level, Table 4.2 shows that the consumption of better-quality food such as fruits and animal products is low even in the food-secure households (according to FIES). By comparison, the severely food-insecure households tend to consume 16, 25, 13 and 15% less fruits, meat, eggs and milk than food-secure households.

Food insecurity is associated with other socio-economic factors. Female-headed households tend to be more food insecure and spend less on food. Food-insecure households also have poorer-quality living standards in terms of access to clean water, quality of their homes, and sanitation facilities. These factors also impact health and nutritional outcomes, particularly of vulnerablepopulations.

**Table 4.1** Food security indicators in rural Tanzania, 2010–2015

|  | Median Annual Food Expenditure Per Capita (US$) | Moderate or Severe Food Insecurity (FIES) (%) | Severe Food Insecurity (FIES) (%) |
|---|---|---|---|
| 2010–2011 | 308 | 31 | 12 |
| 2012–2013 | 412 | 27 | 11 |
| 2014–2015 | 465 | 31 | 14 |

Source: Authors' calculations based on Living Standards Measurement Survey, Tanzania (NBS 2011–2015)

**Table 4.2** Percent of households at different levels of food security consuming fruits and animal products, 2014–2015

|  | Fruit | Meat | Eggs | Milk |
|---|---|---|---|---|
| Food Secure | 66 | 60 | 23 | 38 |
| Moderate or Severe Food Insecurity | 54 | 40 | 11 | 24 |
| Severe Food Insecurity | 50 | 35 | 10 | 23 |

Source: Authors' calculations based on Living Standards Measurement Survey, Tanzania (NBS 2011–2015)

The FIES indicator shows that more than 26% of the world population is estimated to be either moderately or severely food insecure. A large share of the population in low-income countries (62%) and lower-middle-income countries (44%) is food insecure. In sub-Saharan Africa, the rate is 57.7%, with Eastern Africa exceeding at 62.7%. In Asia, 22.8% of the population is reported to be severely or moderately food insecure and it is the highest in South Asia at 34.3%. It is also reported that food insecurity globally is higher among women compared to men, which can be attributed to gender differences in household income, education, and social networks.

There is a 'triple burden of malnutrition' which includes micronutrient deficiencies and obesity in addition to undernourishment. Experts believe that far more people suffer from the first two forms of malnutrition which are underestimated by the usual measures of undernourishment (Gómez et al. 2013). The prevalence of stunting has been decreasing in Africa and Asia. However, more than 30% and 22% of children in Africa and Asia, respectively, were still stunted in 2018. These two regions account for close to 95% of the 149 million stunted children under the age of five worldwide. Stunting is attributed to poor quality of early-life nutrition and repeated infection. Maternal iron deficiency can increase chances of premature birth and of low birth weight and can impede neurological development among infants and young children. About 50% and 40% of the women in their reproductive age were anemic in South Asia and sub-Saharan Africa in 2017 (Ritchie and Roser 2019). On the other hand, there is a steady increase in the incidence of overweight children of less than five years old in Asia and Africa. The two regions accounted for close to 75% of the overweight children under the age of five. Between 2000 and 2018, the percentage of overweight people globally increased from 30.8% to 38.9%. During the same period, in Africa and Asia, the share of overweight people increased by 8.6% and 10%, respectively.

While health and nutritional indicators are closely linked to the consumption of food, farm management for food production and its environmental impact can have adverse results on the health of farming households in particular as well as regionally. Fertilizers and pesticides can be transported by surface runoff and erosion into surface water and leach into the groundwater, and can be transported in the air. The health of farmers and laborers can be exposed while applying pesticides or through incidental exposure by contact with treated vegetation or aerial application. Antle and Pingali (1995) showed that while pesticide use has a positive effect on rice productivity in Philippines, the more toxic chemicals, such as certain insecticides, can have large negative impacts on farmers' health and their productivity (Box 4.3). More recent evidence from the Philippines shows that 50% of the victims of pesticide poisoning were young adults (between twenty-two and thirty-five years old), and even infants and toddlers were exposed through unsafe storage or contaminated clothing (Lu et al. 2010). Poor waste water and excreta management from livestock rearing also leads to public health risks in mixed crop-livestock systems (Lam et al. 2015, 2017). In northern India, stubble burning has contributed to air pollution that has been associated with increased pulmonary diseases as well as permanent lung damage among vulnerable populations such as children (Agarwal et al. 2012; Awasthi et al. 2010). In China, high concentrations of heavy metals such as arsenic have been found in agricultural soils due to industrial emissions, mining, waste water irrigation, fertilizer, and pesticide application (Wei and Yang 2010). This has had an impact on food safety and correlated with high incidence of cancer similar to regions in northwest India (Lu et al. 2015).

Women constitute 55–60% of the agricultural workforce in South Asia and sub-Saharan Africa (World Bank 2019b). The share of female-headed households in developing regions is rising with increasing male migration into urban areas. Yet, women's productivity is often constrained by limited access to land ownership, lack of access to services such as credit, markets, extension services, and household resources, and by restrictive social norms. Daily wages for women are also lower across the developing world. Quantifying the wage differences is difficult since reliable data for farm wages for women are rarely collected. In South Asia and sub-Saharan Africa, women's land ownership is low, between 5 and 20% (Agarwal 2018), and is often confined to marginal lands, resulting in low productivity for female-headed households. According to data from the World Bank, in Tanzania in 2012–2013, maize yields of female-headed farms was 14 % lower than male-headed households. Modern input use was also lower, with 31% female-headed households using hybrid seeds compared to 43% of male-headed farms. Female-owned farms were 35% smaller and fewer owned livestock. With increasing feminization of agriculture, the burden of adaptation and mitigation to climate change will fall increasingly on women. Women are also the primary caregivers to children and play a crucial role for food security. They are actively engaged in food production on the farm, management, preparation, as well as intra-household allocation. Research has shown that women's ownership and control over assets leads to better health outcomes and long-term investments in children's education (Johnson et al. 2016).

## 4.4   Toward More Sustainable Agricultural Systems

Our aim here is to point toward opportunities for moving existing systems in directions that could change—and hopefully improve—their performance in economic, environmental and social dimensions, while recognizing that there will inevitably be tradeoffs as well as synergies. In Sect. 3.4 of Chap. 3 we provided the economic reasons for the existence of tradeoffs across the three dimensions of sustainability. For the very low-external input, largely subsistence systems of the developing world (e.g., Box 3.2), these tradeoffs may be minimal in the environmental dimension, and there are likely to be synergies between higher incomes and the goals of improved food and nutritional security. But as these systems and the economies in which they operate undergo the transition toward more use of modern technology, the environmental and social tradeoffs are likely to intensify—for example, as illustrated by the human health risks associated with increased use of pesticides and other agrichemicals (Box 4.3). Moreover, as low-external input, diversified systems improve their efficiency through better management, economic pressures for increased specialization will arise. For example, the gains from recycling nutrients between crops and livestock can only increase productivity up to a point, beyond which additional grains will require the use of external nutrient inputs in the form of fertilizers or animal feed. These higher efficiencies will in turn encourage more specialization and larger-scale production, the same economic forces that have moved agricultural systems toward larger operations in the industrial world (Box 5.1).

As climate change brings increased economic and weather variability, there will also be a need for systems that are more resilient. The search for strategies to improve the economic as well as environmental performance of crop and livestock systems has led to the concept of 'climate-smart agriculture' (CSA) (Box 4.5).

---

**Box 4.3  Pesticides, Farmer Health, and Sustainable Development**

As farmers increase crop productivity, they have an increased incentive to insure against crop losses from pests and diseases by using chemical pesticides. But many pesticides—most notably, certain insecticides—are known neurotoxins for the target insect pests and also pose acute health risks to humans; they are also harmful to fish and other wildlife. Many pesticides are also known or suspected carcinogens and pose other chronic health risks. Two studies in the 1990s funded by the Rockefeller Foundation carried out analysis of the economic, environmental, and health impacts of pesticide use in rice production in the Philippines and potato production in Ecuador (Pingali and Roger 1994; Crissman et al. 1998). Both studies utilized detailed farm production data combined with observation of the health of farm workers who applied pesticides, typically using backpack sprayers with little protection from exposure to insecticides and fungicides known to pose substantial health risks. Both studies found strong evidence of substantial acute health risks, most notably neurological impairment from exposure to highly toxic insecticides. Both studies also showed there were tradeoffs between reductions in the use of these insecticides and human health. However, better safety procedures for handling and use, and the use of better management practices (often referred to as integrated pest management, or IPM), were found to produce 'win-win' outcomes, thus effectively improving the terms of the economic-health tradeoffs (Fig. 4.10). As a result of this research, the Ecuadorian government instituted efforts to improve pesticide safety, for example, by encouraging farmer field schools to provide training in safe use, and in 2010, banned the most hazardous insecticides. High levels of pesticide use in the Philippines and many other countries continues with minimal regulation and remains a serious health problem. Unfortunately, the use of hazardous pesticides is now spreading into Africa as agricultural productivity has been improving, for example, in irrigated rice-producing areas along the Senegal River in West Africa. A more recent study utilizing data from four African countries with high levels of pesticide use confirmed that the productivity and health impacts of pesticide use found in the earlier studies can be generalized to the African context (Sheahan et al. 2017). Efforts are being made to improve awareness of the problem and develop solutions, for example, the Integrated Production and Pest Management Program supported by the United Nations Food and Agriculture Organization.

*(continued)*

**Box 4.3    (continued)**

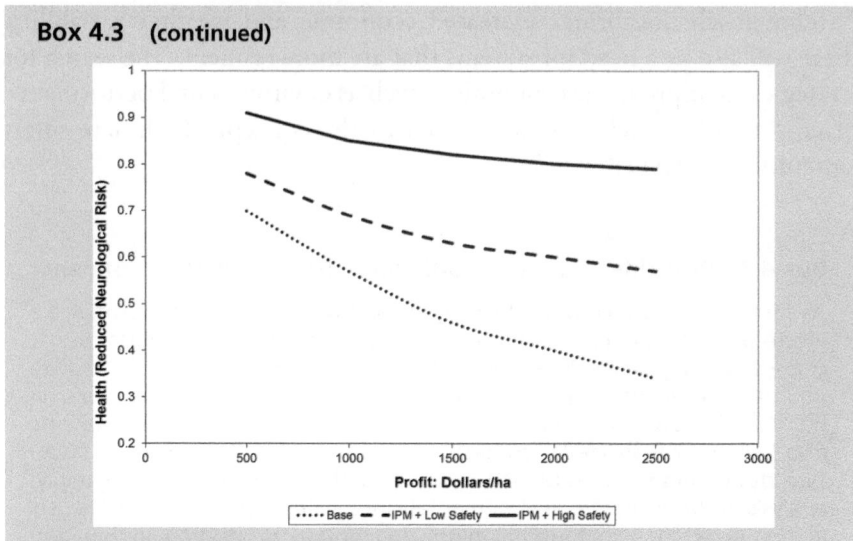

**Fig. 4.11**   Tradeoffs between profitability of Ecuadorian potato production and farm worker health. (Source: Based on data from Crissman et al.
1998. Note: Safety practices include use of protective clothing and appropriate handling and application)

### 4.4.1  Nutrient Management

Adequate soil organic matter and soil nutrients are a critical aspect of
crop management everywhere in the world. But soil management is a
daunting challenge in many parts of the tropics where many soils are
inherently low in organic matter and nutrients, and exhibit other undesirable properties that constrain crop productivity (Fig. 4.11). The soil
nutrient management challenge is particularly acute in regions of Africa
where, due to population pressures, the traditional long fallow system (a
crop grown for one or two seasons, with the land then abandoned for
three or more years) has been replaced with the continuous planting of
crops such as maize with very low external inputs, a practice that rapidly
depletes soil organic matter and soil nutrients.

One solution to this problem for farms that lack access to commercial fertilizers, or the financial ability to purchase it, is to integrate crop production with livestock. In these systems, livestock feed on crop residues and produce manure that is used to fertilize crops to replenish organic matter and soil nutrients. Livestock can also graze on crop residue and their excreta naturally fertilizes land for the next cropping cycle. Similar nutrient cycling is used in small-scale agriculture-aquaculture systems (Box 4.4). These systems have inherent synergies for soil nutrient management at least at small-scale cultivation. Studies from Nigeria, Malawi, and Zambia have shown that crop yield can increase maize production by 118% and cassava production by 26% to 44% (Ndambi et al. 2019). This practice is widely adopted by small-scale agricultural households across the developing world. However, scaling up this system possesses challenges since animal manure can get washed into nearby water bodies leading to surface and groundwater pollution. Large quantities of poorly stored manure can further exacerbate this problem. Many countries including China, the United States, and Brazil have reported water pollution due to untreated discharge of animal waste into water bodies.

**Box 4.4  Nutrient Management in Crop-Livestock and Pond Aquaculture Systems**

Many extremely small farms in Africa and South Asia are unable to maintain soil fertility because they are unable to obtain or afford commercial fertilizers. The result is that soil nutrients are taken up by crops such as maize, rice, and wheat, embodied in grain and essentially 'exported' from the farm system as crops are consumed by family members or marketed. Also, small, subsistence farms that only produce grain and some other subsistence crops provide insufficient protein and other macro- and micronutrients. These small farms also lack the land needed to grow feed crops for livestock. One solution to this problem is to use 'zero-grazing' systems. In these systems, crop residues are collected and fed to animals in pens so that animal manure can be composted and returned to crop fields as fertilizer. Very small farms may not have enough crop residues and may have to supplement the system with fodder crops (such as Napier grass in Kenya) or purchase residues from other farms. In regions with adequate water to maintain ponds for fish aquaculture, such as in Bangladesh and Malawi, similar systems are used to recycle nutrients, except that nutrient-rich pond water is used to irrigate crops (Murshed-E-Jahan et al. 2013)

## 4.4.2   Crop Residue Management

Crop residue management is a good example of an agronomic management practice that is often promoted as part of more environmentally friendly production systems, but also illustrates that there are likely to be tradeoffs between economic and environmental performance. Crop stubble or residues can play an important role in stabilizing the topsoil, prevent erosion, recycle organic matter and nutrients back into the soil, and control weeds. In semi-arid regions, this also helps to retain moisture in the soil by reducing evapotranspiration. Agricultural households with low yields engaged in mixed crop and livestock farming often face a tradeoff between retaining residue on the plot and using it for animal grazing or selling crop residue to other farmers as feed. In sub-Saharan Africa, with low yields, crop residue is usually used as animal feed since households rely on animal products for their own consumption as well as income. It is also important to recognize that crop residue management may require additional labor, and may impact the yields of subsequent crops planted into fields with stubble or residues. For example, incorporation of crop residues into the soil can be difficult for farms dependent on manual labor, and crop residues left on the soil surface may impact soil temperature and crop emergence, either positively or negatively.

---

**Box 4.5   Climate-Smart Agriculture and Sustainable Development**

From a climate justice point of view, the goal of limiting anthropogenic warming of the global climate to 1.5 degrees centigrade is viewed by many as largely the responsibility of the industrialized world where most fossil fuels are consumed. In the developing world, the main concern is arguably to find ways for farmers to adapt to a changing climate, and to improve the resilience of agricultural systems and rural communities to climate variability. To the extent that farmers in the developing world can also help mitigate climate change, for example, through carbon sequestration in soils or biomass, and better livestock management, that also is desirable. Systems with these desirable economic, environmental, and socially desirable attributes are often referred to as 'climate smart.'

(continued)

**Box 4.5  (continued)**

While there is much agreement with these goals—which are similar to goals often attributed to 'conservation' or 'ecological' or 'regenerative' agriculture—it is far less clear how to achieve all of these goals, due to the tradeoffs that are typically encountered in agricultural systems. For example, it is often argued that holistic, integrated, low-input farming practices are climate smart. However, as the examples of zero-grazing systems in Kenya and integrated agriculture-aquaculture systems in Bangladesh have shown, integrated systems can only provide improved economic performance for relatively small farms. Once those efficiencies have been realized through nutrient cycling, the use of external inputs appears to be necessary to improve economic performance, and the usual economic-environment tradeoffs associated with conventional systems are encountered, such as nitrous oxide emissions from fertilizer use and methane emissions from more intensive livestock production. At this point, it appears likely that other kinds of technologies, such as those associated with digital agronomy (Box 5.1), may be needed. Likewise, research shows that other 'low-input' conservation practices are only successful under very specific conditions (Giller et al. 2009). Thus, while 'climate-smart' agriculture embodies desirable goals, such as improved resilience to weather extremes, it is likely that farmers will encounter tradeoffs between these climate-related objectives and other economic, environmental, and social objectives along the pathway to more sustainable development.

In the IGP, increased crop productivity has led to an increase in production of crop residue in the rice-wheat system. At the same time, burning of crop residue has led to adverse impacts on air quality. Crop residue retention in northwest IGP has been recommended as a 'climate-smart' practice. It is an alternative to stubble burning which can also reduce the overall nutrient intensity of the system. Rice crop stubble is rich in nutrients and its incorporation into the soil as mulch can reduce the necessary fertilization rates. Wheat crop residue left in the soil can reduce water loss during warm summer months prior to rice cultivation. Repeated cycles of crop residue retention increase the water-retention capacity of the soil. Direct seeding of rice without puddling as well as direct seeding of the wheat crop in standing rice stubble have been introduced in Punjab and Haryana as CSA. Farm trials have shown that direct seeding into residue can increase wheat yield by 5 tons/hectare in northwest IGP (Singh and Sidhu 2014).

While there are potential synergies, substantial production gains from crop residue retention require careful management of multiple aspects of the cropping system (Giller et al. 2009). Productivity gains from residue retention may be observed after several cropping cycles. Studies have shown that it may take up to three years to observe yield benefits (Gupta et al. 2007). In fact, immediately after adoption, yield is likely to decrease since nutrient release from the residue is a prolonged process. Decomposition of the residue is faster when incorporated into the soil which requires investment in expensive machinery and can be labor intensive. While crop residue retains soil nutrients, it also attracts pests and insects in subsequent cropping cycles which might lead to reduction in crop yield without appropriate management. Often this can lead to increase in chemical insecticide or pesticide application.

### 4.4.3   Rice and Aquaculture

Rice cultivation has been combined with fish farming in the traditional rice-based cropping systems. In China it has been practiced for more than 1700 years. Similar rice-fish systems are also prevalent in Indonesia, Thailand, and Philippines. In South Asia, rice-fish systems are widely practiced in Bangladesh and some regions of adjoining West Bengal in eastern India.

Flooded rice cultivation provides a suitable breeding ground for shallow-water fish such as carp. Rice fields provide shade and attract insects which serve as food for the fish. In turn, the fish release nutrients in the form of waste and oxygenate the water, and their movement facilitates pollination. The fish reduce insect pests such as plant hoppers in the rice plants by 26% (Lansing and Kremer 2011). The rice-fish system reduces the need for additional nutrients and pesticides, contributes to higher rice yield, and is a source of protein in the diet of subsistence agricultural households. Azolla, a type of fern, is also added to the system to increase nitrogen fixation as well as feed for the fish. Integrated rice-fish farming for three years has shown to increase soil nitrogen and phosphorus by 28% and 44%, respectively (Lu and Li 2006). China has the largest area under the rice-fish system, typically found in the mountainous regions in southern China. Rice-fish systems in China also conserve indigenous rice and fish varieties, increasing the genetic diversity of rice farms.

In China, species of the common carp is integrated with rice systems. In Bangladesh, a wider variety of fish species including shrimp and prawns are also widely integrated with rice cultivation. In Bangladesh, alternating rice-fish systems are widely adopted in the flood-prone regions. Rice is grown in the dry season, followed by aquaculture in the subsequent rainy season under natural flooding. Rice-fish cultivation covers about 20% of the area under integrated rice-fish systems in Bangladesh (Dey et al. 2013).

While rice-fish systems help farmers exploit synergies in integrated nutrient management, there are some limitations to the widespread adoption of the system. Rice-fish systems are more labor intensive than rice monocultures. In China, rural-urban migration has led to labor shortages and has contributed to a decline in area under rice-fish systems in recent years. These systems are heavily reliant on rainfall for flooding. Flooded rice cultivation contributes to about half of the methane emissions which is a potent greenhouse gas (Lansing and Kremer 2011). Returns to rice-fish systems in Bangladesh can be higher than rice monocultures, but require careful management and show a high degree of variability in performance. However, integrated systems have lower returns when compared to larger, more specialized continuous fish systems with purchased feed inputs. Further, increasing levels of soil salinity also threaten freshwater fish cultivation in Bangladesh.

## 4.5   Conclusions

The diversity and complexity of the agricultural systems of the developing and transitional regions, in turn, means that the sustainability challenges they face are equally diverse and complex. What they have in common is the challenge of increasing per capita incomes for the growing numbers of people who depend on them, while understanding and balancing the inevitable tradeoffs that will exist with environmental and social impacts. The technological solutions of the Green Revolution have proven not to be a panacea for this very reason, nor have the various schemes that have been promoted more on the basis of desired outcomes than sound science. The lesson we shall carry forward to Chap. 6, where we discuss pathways to sustainable development, is that there is no single, simple solution to these many challenges. Rather, there are many potential solutions. In Chap. 6, we shall discuss the tools and techniques that are being developed to design and implement more sustainable development pathways that are compatible with the goals and capabilities of each system and society.

# References

Agarwal, Bina. 2018. Gender Equality, Food Security and the Sustainable Development Goals. *Current Opinion in Environmental Sustainability, Sustainability Science* 34 (Oct.): 26–32. https://doi.org/10.1016/j.cosust.2018.07.002.

Agarwal, Ravinder, Amit Awasthi, Nirankar Singh, Prabhat Kumar Gupta, and Susheel K. Mittal. 2012. Effects of Exposure to Rice-Crop Residue Burning Smoke on Pulmonary Functions and Oxygen Saturation Level of Human Beings in Patiala (India). *Science of The Total Environment, Special Section— Arsenic in Latin America, An Unrevealed Continent: Occurrence, Health Effects and Mitigation* 429 (July): 161–166. https://doi.org/10.1016/j.scitotenv.2012.03.074.

Aggarwal, P.K., K.K. Talukdar, and Rajesh Mall. 2000. Potential Yields of Rice–Wheat System in the Indo-Gangetic Plains of India. *Rice-Wheat Consortium Paper Series* 10 (Jan.): 16.

Antle, J.M., and P.L. Pingali. 1995. Pesticides, Productivity, and Farmer Health: A Philippine Case Study. In *Impact of Pesticides on Farmer Health and the Rice Environment*, ed. Prabhu L. Pingali and Pierre A. Roger, 361–387. Dordrecht: Springer Netherlands. https://doi.org/10.1007/978-94-011-0647-4_13.

Awasthi, Amit, Nirankar Singh, Susheel Mittal, Prabhat K. Gupta, and Ravinder Agarwal. 2010. Effects of Agriculture Crop Residue Burning on Children and Young on PFTs in North West India. *Science of the Total Environment* 408 (20): 4440–4445. https://doi.org/10.1016/j.scitotenv.2010.06.040.

Barker, Randolph, and Francois Molle. 2004. *Evolution of Irrigation in South and Southeast Asia*. 5. Comprehensive Assessment Research Report. Colombo: Comprehensive Assessment Secretariat. http://www.iwmi.cgiar.org/assessment/files/pdf/publications/ResearchReports/CARR5.pdf.

Bhatt, Rajan, Surinder S. Kukal, Mutiu A. Busari, Sanjay Arora, and Mathura Yadav. 2016. Sustainability Issues on Rice–Wheat Cropping System. *International Soil and Water Conservation Research* 4 (1): 64–74. https://doi.org/10.1016/j.iswcr.2015.12.001.

Chand, Ramesh, P.A. Lakshmi Prasanna, and Aruna Singh. 2011. Farm Size and Productivity: Understanding the Strengths of Smallholders and Improving Their Livelihoods, 7.

Chauhan, Bhagirath S., Gulshan Mahajan, Virender Sardana, Jagadish Timsina, and Mangi L. Jat. 2012. Productivity and Sustainability of the Rice-Wheat Cropping System in the Indo-Gangetic Plains of the Indian Subcontinent: Problems, Opportunities, and Strategies. *Advances in Agronomy* 117 (1): 315–369.

Crissman, C.C., J.M. Antle, and S.M. Capalbo, eds. 1998. *Economic, Environmental and Health Tradeoffs in Agriculture: Pesticides and the Sustainability of Andean Potato Production*, 281 pp. Dordrecht; Boston; London: Kluwer Academic Publishers.

Dey, Madan M., David J. Spielman, A.B.M.M. Haque, M.S. Rahman, and R. Valmonte-Santos. 2013. Change and Diversity in Smallholder Rice–Fish Systems: Recent Evidence and Policy Lessons from Bangladesh. *Food Policy* 43 (Dec.): 108–117. https://doi.org/10.1016/j.foodpol.2013.08.011.

Dhillon, Maninder Singh, Samanpreet Kaur, Anil Sood, and Rajan Aggarwal. 2019. Estimation of Carbon Emissions from Groundwater Pumping in Central Punjab. *Carbon Management* 9 (4): 425–435.

Dixon, John, Aidan Gulliver, and David Gibbon. 2001. *Farming Systems and Poverty*. Accessed September 26, 2019. http://www.fao.org/farmingsystems/mapsregion_en.htm.

Djurfeldt, Agnes Andersson, and Magnus Jirström. 2013. Urbanization and Changes in Farm Size in Sub-Saharan Africa and Asia from a Geographical Perspective, a Review of the Literature. Department of Human Geography, Lund University.

FAO. 2005. Fertilizer Use by Crop in India. Rome: FAO. http://www.fao.org/3/a0257e/A0257E00.htm#TOC.

———. 2010. Asia and Pacific Commission on Agricultural Statistics. http://www.fao.org/fileadmin/templates/ess/documents/meetings_and_workshops/APCAS23/documents_OCT10/APCAS-10-28_-Small_farmers.pdf.

———. 2011. *Irrigation in Southern and Eastern Asia in Figures*. Rome: FAO.

FAO, IFAD, UNICEF, WFP, and WHO. 2019. *The State of Food Security and Nutrition in the World 2019. Safeguarding Against Economic Slowdowns and*

*Downturns.* Rome: FAO. https://www.ifad.org/en/web/knowledge/publication/asset/41220342.

Giller, Ken E., Ernst Witter, Marc Corbeels, and Pablo Tittonell. 2009. Conservation Agriculture and Smallholder Farming in Africa: The Heretics' View. *Field Crops Research* 114 (1): 23–34. https://doi.org/10.1016/j.fcr.2009.06.017.

Gómez, Miguel I., Christopher B. Barrett, Terri Raney, Per Pinstrup-Andersen, Janice Meerman, André Croppenstedt, Brian Carisma, and Brian Thompson. 2013. Post-Green Revolution Food Systems and the Triple Burden of Malnutrition. *Food Policy* 42 (Oct.): 129–138. https://doi.org/10.1016/j.foodpol.2013.06.009.

Grassini, Patricio, and Martin van Ittersum. 2019. Global Yield Gap Atlas. http://www.yieldgap.org/bangladesh.

Gupta, R.K., J.K. Ladha Yadvinder-Singh, Jagmohan Singh Bijay-Singh, Gurpreet Singh, and H. Pathak. 2007. Yield and Phosphorus Transformations in a Rice–Wheat System with Crop Residue and Phosphorus Management. *Soil Science Society of America Journal* 71 (5): 1500–1507. https://doi.org/10.2136/sssaj2006.0325.

Gulati, Ashok, Ranjana Roy, and Siraj Hussain. 2017. *Getting Punjab Agriculture Back on High Growth Path: Sources, Drivers and Policy Lessons.* New Delhi, India: ICRIER. https://icrier.org/pdf/Punjab%20Agriculture%20Report.pdf.

Harvey, Charles F., Christopher H. Swartz, A.B.M. Badruzzaman, Nicole Keon-Blute, Yu Winston, M. Ashraf Ali, Jenny Jay, et al. 2002. Arsenic Mobility and Groundwater Extraction in Bangladesh. *Science* 298 (5598): 1602–1606. https://doi.org/10.1126/science.1076978.

Hobbs, Peter R., and Mahmood Osmanzai. 2011. Important Rainfed Farming Systems of South Asia. In *Rainfed Farming Systems*, ed. Philip Tow, Ian Cooper, Ian Partridge, and Colin Birch, 603–641. Dordrecht: Springer Netherlands. https://doi.org/10.1007/978-1-4020-9132-2_22.

Hou, Fujiang. 2014. Adaptation of Mixed Crop-Livestock Systems in Asia. In *Climate Change Impact and Adaptation in Agricultural Systems.* CABI. http://caoye.lzu.edu.cn/upload/news/N20150323091015.pdf.

ICRISAT. 2015. Meso Level Data for India: 1966–2011, Collected and Compiled under the Project on Village Dynamics in South Asia.

IFPRI. 2014. Farming Systems of Africa. Washington, DC: International Food Policy Research Institute. https://doi.org/10.2499/9780896298460_06.

Jayne, T.S., David Mather, and Elliot Mghenyi. 2010. Principal Challenges Confronting Smallholder Agriculture in Sub-Saharan Africa. *World*

*Development, the Future of Small Farms* 38 (10): 1384–1398. https://doi. org/10.1016/j.worlddev.2010.06.002.

Johnson, Nancy L., Chiara Kovarik, Ruth Meinzen-Dick, Jemimah Njuki, and Agnes Quisumbing. 2016. Gender, Assets, and Agricultural Development: Lessons from Eight Projects. *World Development* 83 (July): 295–311. https:// doi.org/10.1016/j.worlddev.2016.01.009.

Lam, Steven, Hung Nguyen-Viet, Tran Thi Tuyet-Hanh, Huong Nguyen-Mai, and Sherilee Harper. 2015. Evidence for Public Health Risks of Wastewater and Excreta Management Practices in Southeast Asia: A Scoping Review. *International Journal of Environmental Research and Public Health* 12: 12863–12885.

Lam, Steven, Giang Pham, and Hung Nguyen-Viet. 2017. Emerging Health Risks from Agricultural Intensification in Southeast Asia: A Systematic Review. *International Journal of Occupational and Environmental Health* 23 (3): 250–260. https://doi.org/10.1080/10773525.2018.1450923.

Lansing, J. Stephen, and James N. Kremer. 2011. Rice, Fish and the Planet. *Proceedings of the National Academy of Sciences* 108 (50): 19841–19842.

Lowder, Sarah K., Jakob Skoet, and Terri Raney. 2016. The Number, Size, and Distribution of Farms, Smallholder Farms, and Family Farms Worldwide. *World Development* 87 (Nov.): 16–29. https://doi.org/10.1016/j.worlddev. 2015.10.041.

Lu, Jinky Leilanie, Katherine Z. Cosca, and Jocelyn Del mundo. 2010. Trends of Pesticide Exposure and Related Cases in the Philippines. *Journal of Rural Medicine: JRM* 5 (2): 153–164. https://doi.org/10.2185/jrm.5.153.

Lu, Jianbo, and Xia Li. 2006. Review of Rice–Fish-Farming Systems in China— One of the Globally Important Ingenious Agricultural Heritage Systems (GIAHS). *Aquaculture* 260 (1): 106–113. https://doi.org/10.1016/j. aquaculture.2006.05.059.

Lu, Yonglong, Shuai Song, Ruoshi Wang, Zhaoyang Liu, Jing Meng, Andrew J. Sweetman, Alan Jenkins, et al. 2015. Impacts of Soil and Water Pollution on Food Safety and Health Risks in China. *Environment International* 77 (Apr.): 5–15. https://doi.org/10.1016/j.envint.2014.12.010.

Mertz, Ole, Kirsten Halsnæs, Jørgen E. Olesen, and Kjeld Rasmussen. 2009. Adaptation to Climate Change in Developing Countries. *Environmental Management* 43 (5): 743–752. https://doi.org/10.1007/s00267-008-9259-3.

MOSPI. 2019. Ministry of Statistics and Programme Implementation. http:// www.mospi.gov.in/.

Mueller, Barbara. 2017. Arsenic in Groundwater in the Southern Lowlands of Nepal and Its Mitigation Options: A Review. *Environmental Reviews* 25 (3): 296–305. https://doi.org/10.1139/er-2016-0068.

Murshed-E-Jahan, K., C. Crissman, and J. Antle. 2013. Economic and Social Impacts of Integrated Aquaculture-Agriculture Technologies in Bangladesh. CGIAR Research Program on Aquatic Agricultural Systems, Penang, Malaysia. Workshop Report AAS-2013-02. https://www.worldfishcenter.org/content/economic-and-social-impacts-integrated-aquaculture-agriculture-technologies-bangladesh.

National Bureau of Statistics (NBS) [Tanzania]. 2011–2015. Tanzania National Panel Survey Report (NPS)—Wave 2, 2010–2011, Wave 3, 2012–2013, Wave 4, 2014–2015. Dar es Salaam, Tanzania: NBS. www.nbs.go.tz.

Ndambi, Oghaiki Asaah, David Everett Pelster, Jesse Omondi Owino, Fridtjof de Buisonjé, and Theun Vellinga. 2019. Manure Management Practices and Policies in Sub-Saharan Africa: Implications on Manure Quality as a Fertilizer. *Frontiers in Sustainable Food Systems* 3. https://doi.org/10.3389/fsufs.2019.00029.

Nelson, Gerald C., Hugo Valin, Ronald D. Sands, Petr Havlík, Helal Ahammad, Delphine Deryng, Joshua Elliott, et al. 2014. Climate Change Effects on Agriculture: Economic Responses to Biophysical Shocks. *Proceedings of the National Academy of Sciences* 111 (9): 3274–3279. https://doi.org/10.1073/pnas.1222465110.

OECD-FAO. 2017. OECD-FAO Agricultural Outlook 2016–2026. Paris: OECD Publishing. https://www.oecd-ilibrary.org/docserver/agr_outlook-2017-en.pdf?expires=1567972230&id=id&accname=guest&checksum=EF529BF8F2C3EEB8D30ABBBD30FF908C.

Parthasarathy Rao, P., and P.S. Birthal. 2008. *Livestock in Mixed Farming Systems in South Asia.* ICRISAT. https://cgspace.cgiar.org/handle/10568/400.

Pingali, P.L., and P.A. Roger, eds. 1994. *Impacts of Pesticides on Farmer Health and the Rice Ecosystem.* Boston: Kluwer Academic Publishers.

Pingali, P.L., and M. Shah. 2008. Policy Re-directions for Sustainable Resource Use. *Journal of Crop Production* 3 (2): 103–118.

Pingali, Prabhu, Kate Schneider, and Monika Zurek. 2014. Poverty, Agriculture and the Environment: The Case of Sub-Saharan Africa. In *Marginality: Addressing the Nexus of Poverty, Exclusion and Ecology*, ed. Joachim von Braun and Franz W. Gatzweiler, 151–168. Dordrecht: Springer Netherlands. https://doi.org/10.1007/978-94-007-7061-4_10.

Ritchie, Hannah, and Max Roser. 2019. Micronutrient Deficiency. OurWorldInData.Org. https://ourworldindata.org/micronutrient-deficiency.

Robinson, T.P., P.K. Thornton, G. Franceschini, R.L. Kruska, F. Chiozza, A. Notenbaert, G. Cecchi, et al. 2011. *Global Livestock Production Systems* (152 pp). Rome, Food and Agriculture Organization of the United Nations (FAO) and International Livestock Research Institute (ILRI).

Roser, Max, and Esteban Ortiz-Ospina. 2013. Global Extreme Poverty. Our World in Data, May 25. https://ourworldindata.org/extreme-poverty.

Satoh, Yusuke, Taher Kahil, Edward Byers, Peter Burek, Günther Fischer, Sylvia Tramberend, Peter Greve, et al. 2017. Multi-Model and Multi-Scenario Assessments of Asian Water Futures: The Water Futures and Solutions (WFaS) Initiative. *Earth's Future* 5 (7): 823–852. https://doi.org/10.1002/2016EF000503.

Sheahan, Megan, and Christopher B. Barrett. 2017. Ten Striking Facts about Agricultural Input Use in Sub-Saharan Africa. *Food Policy, Agriculture in Africa—Telling Myths from Facts* 67 (Feb.): 12–25. https://doi.org/10.1016/j.foodpol.2016.09.010.

Sheahan, Megan, Christopher B. Barrett, and Casey Goldvale. 2017. Human Health and Pesticide Use in Sub-Saharan Africa. *Agricultural Economics* 48: 27–41.

Singh, Yadvinder, and H.S. Sidhu. 2014. Management of Cereal Crop Residues for Sustainable Rice-Wheat Production System in the Indo-Gangetic Plains of India. *Proceedings of the Indian National Science Academy* 80 (1): 95–114.

Syuaib, M. Faiz. 2016. Sustainable Agriculture in Indonesia: Facts and Challenges to Keep. Growing in Harmony with Environment. *Agricultural Engineering International: The CIGR e-Journal* 18 (June): 170–184.

Tegemeo Institute. 2019. Tegemeo Agricultural Monitoring and Policy Analysis Project Data. https://www.tegemeo.org/index.php/resources/data.html

Upreti, Bishnu Raj, and Yamuna Ghale Upreti. 2002. Factors Leading to Agro-Biodiversity Loss in Developing Countries: The Case of Nepal. *Biodiversity & Conservation* 11 (9): 1607–1621. https://doi.org/10.1023/A:1016862200156.

Valdivia, Roberto O., John M. Antle, and Jetse J. Stoorvogel. 2017. Designing and Evaluating Sustainable Development Pathways for Semi-Subsistence Crop-Livestock Systems: Lessons from Kenya. *Agricultural Economics* 48 (S1): 11–26. https://doi.org/10.1111/agec.12383.

Virk, Hardev. 2018. A Preliminary Report on Groundwater Contamination of Majha Belt of Punjab Due to Arsenic and Selenium Heavy Metals.

Wang, Jieyong, Ziwen Zhang, and Yansui Liu. 2018. Spatial Shifts in Grain Production Increases in China and Implications for Food Security. *Land Use Policy, Land Use and Rural Sustainability in China* 74 (May): 204–213. https://doi.org/10.1016/j.landusepol.2017.11.037.

Wei, Binggan, and Linsheng Yang. 2010. A Review of Heavy Metal Contaminations in Urban Soils, Urban Road Dusts and Agricultural Soils from China. *Microchemical Journal* 94 (2): 99–107. https://doi.org/10.1016/j.microc.2009.09.014.

World Bank. 2019a. Employment in Agriculture. https://data.worldbank.org/indicator/SL.AGR.EMPL.ZS?locations=8S-ZF-CN.

———. 2019b. Employment in Agriculture, Female. https://data.worldbank.org/indicator/SL.AGR.EMPL.FE.ZS?locations=8S-ZG&view=chart.

———. 2019c. Fertilizer Consumption (Kilograms per Hectare of Arable Land) | Data. https://data.worldbank.org/indicator/AG.CON.FERT.ZS?locations=ZG-TZ-ET-MW-8S.

Wu, Yiyun, Xican Xi, Xin Tang, Deming Luo, Baojing Gu, Shu Kee Lam, Peter M. Vitousek, and Deli Chen. 2018. Policy Distortions, Farm Size, and the Overuse of Agricultural Chemicals in China. *Proceedings of the National Academy of Sciences* 115 (27): 7010–7015. https://doi.org/10.1073/pnas.1806645115.

Xin, Liangjie, Xiubin Li, and Minghong Tan. 2012. Temporal and Regional Variations of China's Fertilizer Consumption by Crops during 1998–2008. *Journal of Geographical Sciences* 22 (4): 643–652. https://doi.org/10.1007/s11442-012-0953-y.

You, Liangzhi, Claudia Ringler, Ulrike Wood-Sichra, Richard Robertson, Stanley Wood, Tingju Zhu, Gerald Nelson, Zhe Guo, and Yan Sun. 2011. What Is the Irrigation Potential for Africa? A Combined Biophysical and Socioeconomic Approach. *Food Policy, Between the Global and the Local, the Material and the Normative: Power struggles in India's Agrifood System* 36 (6): 770–782. https://doi.org/10.1016/j.foodpol.2011.09.001.

Zhang, Jianhua. 2011. China's Success in Increasing per Capita Food Production. *Journal of Experimental Botany* 62 (11): 3707–3711. https://doi.org/10.1093/jxb/err132.

# 5

# Challenges of Sustainable Agricultural Development in High-Income Countries

## 5.1 Introduction

This chapter extends the discussion of sustainable development challenges to the agricultural systems of the high-income, industrialized countries that also have nationally and globally important agricultural sectors. These countries include: Canada and the United States in North America; Argentina, Brazil, and Chile in South America; the United Kingdom and the larger Western and Central European countries and Russia; South Korea and Japan in East Asia; and Australia. Some of these countries such as Brazil and Russia could be considered transitional based on aggregate per capita incomes, but we include them here because they are large industrial countries that also play an important role in global agricultural markets. China of course now also is a major industrial producer but was included in Chap. 4 due to its persistent large, poor rural populations and predominantly small-scale agriculture. In addition to having relatively high incomes, these countries all have a relatively small share of labor employed in agriculture, and agriculture represents a relatively small share of GDP, and except for Brazil most have temperate climates. Even though agriculture represents a small share of GDP in these

© The Author(s) 2020
J. M. Antle, S. Ray, *Sustainable Agricultural Development*, Palgrave Studies in Agricultural Economics and Food Policy, https://doi.org/10.1007/978-3-030-34599-0_5

high-income countries, these countries' agricultures produce a disproportionate share of the world's output and about 50% of global trade in agricultural commodities, while representing only about 15% of global population (Beckman et al. 2017; World Bank Data). Moreover, these countries' agricultures are experiencing continuing growth and change, driven by their integration into global food systems, by demand-side drivers such as demographics and food preferences and by supply-side drivers such as technology and policy.

Despite their many similarities, this is nevertheless a diverse group of countries in terms of agriculture. Some industrialized countries are very large and land abundant and export a large share of their major agricultural products, whereas some are relatively small and land scarce with high human population densities. There are also substantial differences on the demand-side associated with demographics and culture, as well as supply-side differences such as farm size, technology, and policy. Although most of these countries are located in the Northern Hemisphere or in temperate regions of the Southern Hemisphere, there are large physical differences in soils and climate, as well as substantial differences in institutions and domestic and trade policies. Indeed, within some of the larger countries such as Brazil and the United States, there is as much or more agricultural diversity within as between countries, and as much or more heterogeneity within farming systems as between the systems of different countries.

Thus, as with the low-income and transitional countries discussed in Chap. 4, among the industrialized countries we find that the diversity and heterogeneity of agriculture leads to a complex picture of economic, environmental, and social dimensions of sustainability. While the principles of sustainability we discussed in Chaps. 2 and 3 apply globally, the sustainability challenges facing the agricultures of the industrialized world are quite different from those of the lowest income regions but have some similarities to the transitional regions. Whereas in the low-income countries extreme poverty and food and nutritional security are priorities for rural and poor urban households, in the industrialized countries, farm household incomes often equal or exceed non-farm incomes. This is not to say that poverty and food insecurity have been eradicated in the higher-income world—indeed, in rural regions as well as in urban areas of many industrial countries, one finds persistent pockets of poverty and food insecurity—but these problems are not typical of the households operating commercial agricultural enterprises. Instead, financial insecurity is an issue for many commercial-scale producers, particularly for those who borrow to finance the purchase of land, equipment, and structures and for operating capital. This financial insecurity is also due to high commodity price instability and policy risk, as illustrated by the vagaries of recent trade policy in the United States and other parts of the world.

As we discussed in Chap. 4, another characteristic of farms in the low-income regions of the world is their very low use of externally purchased inputs, particularly improved seeds, fertilizers, pesticides, and fossil fuels for crop production, and purchased feed and drugs such as antibiotics for livestock production. Thus, apart from issues associated with land use such as soil erosion on steeply sloped fields and clearing of forests by logging or burning, the agricultural systems of the low-income countries have relatively low off-farm environmental impact. In contrast, industrial agricultural systems tend to be much more intensive, utilizing large amounts of synthetic fertilizers and pesticides, drugs to manage the health of livestock produced in confinement, large amounts of animal waste, genetically modified crops and livestock, and fossil fuels to power machinery. Thus off-farm environmental and food quality and safety are major issues for industrial agriculture.

There are also important but different social issues associated with developing and industrialized agricultures. Whereas many agricultural regions of low-income countries are challenged by high population densities and the need to facilitate migration from rural to urban areas or from agriculture to non-agricultural activities in rural areas, the mechanization and increasing automation of large-scale industrial agriculture tends to be associated with low population densities in rural areas. Thus, social challenges for rural areas in industrial countries tend to be associated with declining and aging populations—how to maintain viable communities with essential social, health, and educational services.

In the next section we review the major agricultural systems of the major industrialized countries and then review the sustainability indicators relevant to these systems. Next we discuss the current status of these systems in the three dimensions of sustainability.

## 5.2    Major Systems and Characteristics

Figures 3.2 and 3.5 summarize the type and location of major crop-based and livestock-based systems of the world which span low-income, transitional, and industrialized countries—the major cereal grain and oilseed commodities used for both human consumption and animal feed, fruits and nuts, vegetables, sugar, tea, coffee, beef, pork, chicken and dairy, and nonfood fiber and fuel crops, as well as a large number of specialty crops such as herbs and spices, hops for beer, wine grapes, and grass seed. As we noted in Chap. 3, major differences between the systems in lower-income countries and the higher-income transitional and industrialized countries can be seen in farm size, degree of specialization, use of technology, and integration into national and global markets and supply chains. Our example of dryland wheat farms in the United States (Box 3.1) illustrates the prototypical industrial farm, a true monoculture (a single crop) produced with capital-intensive technology, selling one main product into a global market.

### 5.2.1    Farm Size and Ownership

A fundamental characteristic of farm households we identified in Chap. 3 is farm size, measured by land area cultivated for crop farms, and by number of animals for livestock and dairy producers. A trend among almost all of the industrialized countries is toward a smaller number of larger farms producing a larger share of farm output (Bokusheva and Kimura 2016). This trend is explained by various factors, but a key factor is the better financial performance of larger farms and their capability to provide an income comparable to non-agricultural sectors to highly skilled farm managers. Other important factors include labor scarcity, which encourages large-scale, capital-intensive technologies such as automation and information technology. Another feature of agriculture in many industrialized countries is an aging farm household population. Other economic factors favoring larger farms include economies of scale in purchasing inputs and marketing output and in utilizing specialized management services for nutrient and pest management and for financial and risk management (MacDonald et al. 2013).

**Table 5.1** Acreage of field crop farms in the United States, 1987, 1997, and 2007

| Commodity | Midpoint acreage | | |
| --- | --- | --- | --- |
| | 1987 | 1997 | 2007 |
| Corn | 200 | 350 | 600 |
| Cotton | 450 | 800 | 1090 |
| Rice | 295 | 494 | 700 |
| Soybeans | 243 | 380 | 490 |
| Wheat | 404 | 693 | 910 |
| Hay | 120 | 140 | 160 |

Source: MacDonald et al. (2013)

Note: Midpoint acreages are the enterprise farm size, in harvested acres, at which half of all harvested acres are on larger enterprises, and half are on smaller enterprises

Data from the Unites States illustrate these trends and what they mean for the current and likely future farm size distribution. Table 5.1 shows that the average number of acres of major commodity crops per farm doubled or tripled over the past three decades. Figure 5.1 shows that crop agriculture is now dominated by a relatively small number of very large crop farms that produce most of the output; yet, there is also a large number of very small farms that contribute a relatively small share of total output. Most of these small farms are part-time, often 'hobby farms,' and produce a small and declining share of total output. The number and importance of middle-sized commercially oriented farms, sometimes referred to as the 'agriculture in the middle,' has declined. We focus most of our attention on the larger commercially oriented farms that produce most of the output and have the most economic and environmental impacts. Yet, it is important to keep in mind that the smaller farms do play a role in supporting rural communities and do influence the rural landscape, especially in the peri-urban areas where the smaller farms tend to predominate.

Despite the increasing scale of commercial farms, family farms (with the principal operator and family members owning 50% or more of assets) continue to play a dominant role in US agriculture and the agricultures of most other industrialized countries. In 2015, these farms accounted for 99% of US farms and 89% of production. About 90% of farms in the United States were small family operations that accounted for 24% of the value of production, whereas large-scale family farms were only 2.9% of US farms but contributed 42% of total production; non-family farms accounted for only 11% of agricultural production.

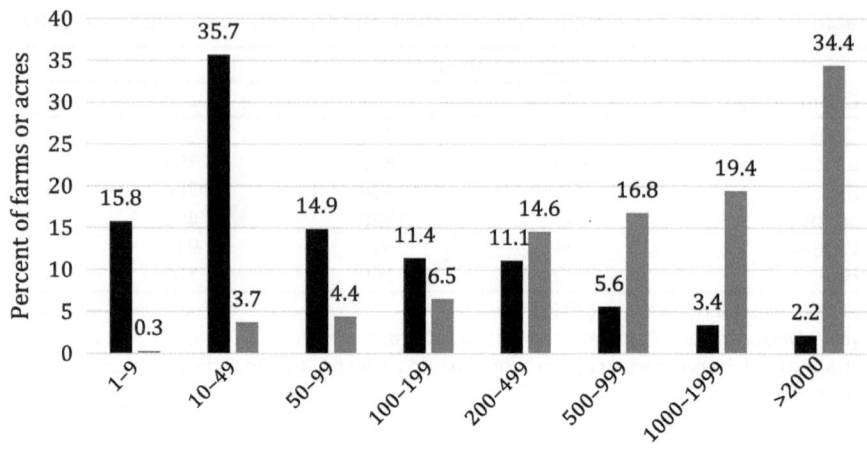

**Fig. 5.1** The size distribution of crop farms in the United States, 2011. (Note: Mean farm size (total cropland divided by total farms with cropland) is 234 acres. Half of all farms have less than 45 acres (the median), and half have more. Half of all acres are on farms with less than 1100 acres (the midpoint acreage), and half are on farms with more. Farm size is defined according to the cropland the farm operates—that is, the cropland it owns, plus any that it rents, minus any rented to others. Source: MacDonald and McBride (2009))

Similar farm size trends and dominance of family farms are found in most higher-income countries, but there are substantial differences in farm size primarily due to the relative abundance or scarcity of land and labor. As we discuss in Chap. 6, the availability of land has a strong influence on the types of crops and livestock produced and the technologies that are used. In Japan, where population density is high and land scarce, average farm size remains around 2 hectares, whereas in land abundant countries like the United States, average farm size is over 200 hectares and many commercial crop farms are an order of magnitude larger, as shown in Table 5.1. Most other industrialized countries fall between these two extremes (Bokusheva and Kimura 2016).

**Table 5.2** Scale of livestock operation in the United States, 1987, 1997, and 2007

| | Midpoints | | |
|---|---|---|---|
| Commodity | 1987 | 1997 | 2007 |
| Broilers | 300,000 | 480,000 | 681,600 |
| Hogs | 1200 | 11,000 | 30,000 |
| Fattened cattle | 17,532 | 38,000 | 35,000 |
| Cattle (<500 lbs) | 50 | 65 | 128 |
| Milk cow | 80 | 140 | 570 |

Source: MacDonald et al. (2013)
Note: The midpoint is defined as the enterprise size, in number of head, at which half of animals are on larger enterprises and half are on smaller enterprises.

In 1987, the midpoint dairy herd size was 80 cows; by 2007, it was 570 cows in the United States. The change in hogs was even more striking, from 1200 hogs removed in a year to 30,000. Farm consolidation was widespread over this period. The midpoint number of head sold for fed cattle doubled between 1987 and 2007, while those for broilers and cow-calf operations (cattle, less than 500 pounds) more than doubled (MacDonald et al. 2013; Table 5.2).

## 5.2.2 Specialization and Technology

One of eighteenth-century economist Adam Smith's insights was that productivity gains are made possible through specialization, and this is as true in agriculture as in manufacturing. While agricultural specialization is also determined by physical (soils, climate) and biological (genetics, pests, and diseases) factors, the trend toward larger and more specialized commercial farms is clearly driven by this basic economic truth. We see this across most types of crops and livestock production systems. In addition to increasing specialization of individual farms, there has been regional (within-country) specialization through the relocation of production. A key element of this specialization is the separation of food crop, feed crop, and livestock production. For example, in the United States, as farms became more specialized, Corn Belt states concentrated more heavily on the production and sale of feed crops, and livestock production moved from the Corn Belt to Mountain, Southeastern, and Southern Plains states. In 1950, crops accounted for less than a third of cash receipts from farming in the four Corn Belt states of Illinois, Indiana, Iowa, and Ohio. Most cash receipts came from sales of livestock fed on the crops grown on those farms. Sixty years later, livestock sales were no longer dominant: crops accounted for 68% (MacDonald et al. 2013).

The foundation of the agricultural revolution of the twentieth century was the improvement of crops and livestock genetics through breeding research that selected and crossbred crops and livestock with desirable traits. Modern genetic techniques have made it possible to greatly accelerate this process by allowing scientists to directly alter genes to introduce desirable traits such as herbicide resistance or drought tolerance. Since 1985, thousands of genetically modified organisms (GMOs) have been developed and used in production of many major crop and livestock species (Freedman 2013; James 2013; ISAAA 2018). Despite their rapidly growing importance worldwide, the use of crops and animals with genes modified through 'genetic engineering' methods is opposed by some people based on ethical, health, and safety concerns. Indeed, some countries have substantially restricted the use or sale of 'genetically modified organisms (GMOs),' and many certification programs for 'organic' or 'sustainable' labeling prohibit them. A major concern both for producers and consumers is the ability to prevent unintended spread of GMOs in the environment and to be able to trace GMOs through supply chains so that people know whether products contain them. Nevertheless, the scientific consensus is that GMOs are generally safe as long as their properties are carefully evaluated for possible health risks such as allergenicity (World Health Organization 2014).

Increases in scale and specialization are also linked closely to advances in mechanical technology, including tractors guided by global positioning systems and automation and digital technology for the so-called precision agriculture for planting, nutrient, and pest and water management

**Table 5.3** Changes in planting and harvesting machinery in field crops in the United States, 1970–2010

| Year | Planting efficiency Technology | Outcome (acres/day) | Harvesting efficiency Technology | Outcome (bu./day) |
|---|---|---|---|---|
| 1970 | 4 rows @ 2 mph | 40 | 4 rows, 12 hrs/day | 4000 |
| 2005 | 16 rows @ 6 mph | 420 | 12 rows, 12 hrs/day | 30,000 |
| 2010 | 36 rows @ 6 mph | 945 | 16 rows, 12 hrs/day | 50,000 |

Source: Bechdol et al. (2010)

(see Box 5.1). For example, in field crop production, from 1970 to 2010, planting efficiency (acres planted per day) increased by over 2000% and harvesting efficiency (amount harvested per day) increased by over 1000% in US agriculture (Table 5.3).

Digital technology also is playing an increasingly important role in providing weather forecasts and in managing financial and marketing decisions by providing price data and allowing farmers to utilize sophisticated financial market instruments such as futures and options contracts. The high cost of these technologies and the high level of management skill required for their use, in turn, increase the economic efficiency of large farms that can spread fixed physical and human capital costs over more units of production. These technological advances also encourage the use of contracted specialized services. Large-scale grain farms have long employed contracted services for grain harvest. By replacing the farmer's site-specific knowledge of soil and climate conditions with information obtained from satellite data and sensors located on machinery and aerial devices, it is possible to outsource many management decisions and services including seed selection and nutrient, pest, and water management. Technology also facilitated the increasing scale in livestock operations, particularly in hog, dairy, fed cattle, and poultry production (MacDonald and McBride 2009). Close integration with processors through long-term production contracts and relocation and increased size of operations also encouraged livestock consolidation.

---

**Box 5.1   Digital Agronomy and Sustainable Intensification of Crop Agriculture**

Agronomy is the science of crop production. Until recently, crop management has depended on the farmer's intimate knowledge of the land and climate. Agronomic science improved crop management through classical experimentation to improve crop genetics and management practices needed to realize those improvements in the field. These traditional methods are now being replaced by digital and geospatial technologies that integrate sensors, computer-based analytics, and automation to monitor, assess, and manage soil and climatic and genetic resources with appropriate practices and inputs. Many scientists view the goal of these technologies to be *sustainable intensification*, that is, to achieve the dual goals of increasing productivity while also improving environmental performance by more precisely applying seeds and nutrients according to site-specific soil and climate conditions. Until now, however, it is unclear whether the use of so-called precision agriculture has been able to achieve sustainable intensification. For example, many farmers utilize precision technology to apply more nitrogen to low-yielding portions of fields in the hope of increasing yields, rather than less nitrogen to avoid fertilizer losses through leaching and runoff of nitrogen that crops cannot use. This tendency is compounded by apparent conflicts between farmers' goals (e.g., to maximize profit) and the objective of input suppliers (to maximize input sales). Thus, ironically, precision management tools may result in lower economic and environmental sustainability if not used appropriately. By better integrating remotely sensed data on crop yield variation with highly detailed soils and weather data, fertilizer application 'tactics' could achieve high yields with lower environmental impact and better economic outcomes for farmers (Basso and Antle 2020). For further discussion of the role of sustainable intensification in sustainable agricultural development, see Thematic Group on Sustainable Agriculture and Food Systems (2013).

---

## 5.2.3   Market and Supply Chain Linkages

Most commercial agricultural producers of major commodities are linked to national or global supply chains. These supply chains connect producers to consumers through various combinations of transportation, storage, processing, and wholesale and retail food businesses. For example, the wheat produced in the US Pacific Northwest (Box 3.1) is usually transported to grain elevators for storage in the region where it is produced and may then be transported by train to Portland, Oregon where it is loaded on ships, transported to Japan, off-loaded, and transported to grain mills, with the wheat flour then packaged and marketed to bakers or to retailers.

Similarly, corn and soybean producers in the US Midwest either store grain on-farm or ship grain to storage facilities, which in turn link to rail and river transport. Some are processed regionally, both for food, feed and biofuels; much of Midwestern grain is transported via the Mississippi river and then shipped to international destinations, to be in turn used in other countries' food, feed, or biofuel industries. Similar storage and transportation systems are utilized in other major commodity-exporting countries and play a key role in facilitating or limiting their participation in global markets. For example, one of the significant constraints to Brazil's development as a major soybean exporter has been the development of adequate transportation from major producing regions such as the state of Mato Grosso, located approximately 1000 miles from the Atlantic coast and blocked from transport to the Pacific coast by the lack of adequate rail or road transport over the Andes mountains. The consequence is that transport cost for soybeans to major international destinations, such as Shanghai and China, is approximately 50% higher from Mato Grosso than from the state of Iowa in the United States (Salin 2018).

A key factor in food supply chain development over the past decade has been the growing demand from consumers, and thus from food processors, wholesalers, and retailers, for greater transparency in food origin, production methods, and other credible quality attributes such as use of pesticides and GMOs. As we discuss in Chap. 6, these quality attributes play an important role in designing and achieving sustainable agricultural development pathways. This is another factor that tends to encourage smaller numbers of large, highly specialized producers in a relatively small geographic region, because the cost of creating and transmitting quality information from producers through the supply chain to consumers is lower for a small number of larger, geographically concentrated producers than for a larger number of small producers, or for producers that are spatially dispersed.

## 5.3 Key Sustainability Indicators and Current Status of Industrial Systems

Tables 3.1, 3.2 and 3.3 summarize many of the widely used sustainability indicators at the household farming system level. As we discussed in Chap. 3, for each system, there will be a subset of these indicators that is most relevant. Moreover, as a practical matter, it is useful to identify a small number of quantifiable indicators in the three dimensions of sustainability. For industrial systems, in addition to key economic indicators (farm and household income, financial condition and technology), industrial agricultural systems are associated with the full complement of environmental impacts, including the effects of land use change and the potential impacts of intensive production systems on soil, water, air, and biodiversity. In the social dimension, food insecurity and undernutrition are not typically associated with households engaged in commercial agriculture, although obesity can be an issue. But other important health and safety issues are associated with commercial systems, including farmworker safety and mental health challenges associated with financial stress. Industrial systems are also associated with various off-farm social issues. In many countries, farm consolidation and capital-intensive, increasingly automated systems are reducing the demand for farm labor and contributing to a declining and aging rural population in areas that were historically sparsely populated. These trends contribute to the decline of local communities and increase the cost of educational, medical, and social services. Industrial systems also are associated with broader concerns for the humane treatment of livestock, for example, limiting the use of confinement, maintaining good animal health, and humane slaughter. Other market-related issues such as supporting local family farms are also of interest to some people.

When we aggregate across systems to the regional (subnational), national, or global scales, it is also useful to focus on some key quantifiable indicators. For example, the FABLE (2019) research project selected the following three 'pillars' and indicators for national analysis of agricultural system and wider food system sustainability, similar to the three dimensions of sustainability we identified in Chap. 3:

Pillar 1: Efficient and resilient agricultural systems; indicators: yields, food loss, greenhouse gas emissions, water use, and nitrogen and phosphorous pollution.

Pillar 2: Conservation and restoration of biodiversity; indicators: deforestation, land use change, and biodiversity conservation.

Pillar 3: Food security and healthy diets; indicators: hunger, dietary disease risk, and food waste.

Similarly, the Organisation for Economic Co-operation and Development (OECD) focused its national-level agri-environmental indicator data and analysis for OECD countries on the indicators presented in Table 5.4. These indicators also include agricultural, nutrient, pesticide, water, and energy use.

**Table 5.4** OECD agri-environmental indicators, 2008–2010

| Percentage of OECD primary agriculture in total | OECD average | Range of values (minimum to maximum) |
|---|---|---|
| GDP | 2.6% | 0.3–9.2% |
| Land area | 36% | 3–72% |
| Certified organic farm area as a share of total agricultural area | 1.9% | 0.01–15.6% |
| *Nutrient balances (surpluses and deficits):* | | |
| Nitrogen, kg per hectare of agricultural land | 63 kg/ha | 1–228 kg/ha |
| Phosphorous, kg per hectare of agricultural land | 6 kg/ha | −10–49 kg/ha |
| Pesticide sales | 70% | 65–80% |
| Energy consumption | 1.6% | 0.4–6.3% |
| Water withdrawal | 44% | 0.2–89% |
| *Irrigated land area share in total agricultural area:* | 4% | 0.4–54% |
| Nitrates in surface water | – | 33–82% |
| Nitrates in groundwater | – | 1–34% |
| Nitrates in coastal water | – | 35–78% |
| Phosphorous in surface water | – | 17–70% |
| Phosphorous in coastal water | – | 23–50% |
| Pesticides in coastal water | – | 0–75% |
| Pesticides in groundwater | – | 0–25% |
| Ammonia emissions | 91% | 82–98% |
| *Greenhouse gas emissions of which:* | 8% | 2–46% |
| Nitrous oxide emissions | 75% | – |
| Methane emissions | 38% | – |
| *Share of OECD methyl bromide used in the world:* | | |
| Ozone-depleting products | 5% | – |
| Methyl bromide use | 46% | – |

Source: OECD (2013)

### 5.3.1   Economic Status

Higher profitability of larger farms will continue to drive the trend to larger and more specialized crop and livestock farms. Domestic and trade policies will continue to play an important role in farm profitability and farm incomes, even though there has been a trend toward lower levels of agricultural subsidies in the industrialized countries over the past several decades (FAO and OECD 2019).

A key question facing agriculture globally, and particularly for the industrialized agricultural sectors linked through global markets, is the long-term trend in prices, productivity, and costs of production. Some analysts argued in the early 2000s that the long-term historical downward real price trend for major commodities might be reversed in the early part of the twenty-first century due to slowing agricultural productivity growth and increasing global population and incomes (Porter et al. 2014). However, after price spikes in the 2008–2012 period, most major commodity prices declined substantially although they have remained somewhat above the levels prior to 2008–2012 in real terms. The OECD is now projecting that real prices will remain at or below the long-term trend for the next decade as productivity improvements continue to outpace demand growth, albeit with likely continued price volatility. OECD projects that agricultural production will grow by 15% over the coming decade, despite little change in global agricultural land use. Growth in crop production will be due largely to increased productivity driven by ongoing technological innovation, particularly through genetic improvement and increased efficiency from mechanical innovations and the growing use of digital technology and automation. Livestock production will increase due to larger herds, driven by growing incomes in the middle-income countries. Aquaculture will provide most of the growth in fish production and consumption.

Continuing global population growth will continue to expand demand for food, feed, fuel, and other industrial purposes. Much of the additional growth in basic food commodities will come from regions with high population growth, mainly sub-Saharan Africa, India, and the Middle East and North Africa. Per capita consumption of staple foods is expected to grow little or decline, as demand is satiated for most of the world's population. Meat demand growth is expected to be strong in the Americas. Demand growth in lower-income regions such as sub-Saharan Africa will increase with continued high population growth but is likely to be constrained by low per capita income growth. Fresh dairy products will satisfy much of the protein demand growth in South Asia. Per capita consumption of sugar and vegetable oils is expected to increase along with urbanization and the shift toward more highly processed and convenience foods.

Income growth, particularly in middle-income countries undergoing the 'dietary transition' from vegetable-based protein to animal-based protein, will continue to drive an increasing demand for animal food products. This will in turn drive growth in the demand for animal feed, with the result that growth in feed use of cereals is expected to exceed the expansion of food use over the coming decade. Biofuels formed a major source of crop demand growth between 2000 and 2015 but is not expected to grow as rapidly in the next decade with persistent low fossil fuel prices and no major changes in policies to increase use.

Economic risks for industrial agriculture remain high and are likely to increase in the next decade. On the demand side, they include evolving consumer preferences in middle- and high-income countries, including a growing demand for certifiably safe, healthy food produced sustainably. Consumer and government policy responses to the obesity epidemic also represent major uncertainties. Uncertainty in domestic subsidy policies is being compounded by erratic changes in trade policy among major producing and consuming countries, such as Brazil, China, and the United States. The escalation of trade disputes and the increasing use of tariffs as well as nontariff barriers are likely to make substantial changes in trade flows and have large impacts in international and domestic markets.

On the supply side, uncertainties include the spread of crop diseases such as Fusarium wilt, which threatens global banana production, and animal diseases such as African Swine Fever that has spread from Africa to Asia, growing resistance to antimicrobial substances, regulation of biotechnology, and increasing risks of extreme climate events associated with ongoing climate change. Additionally, new innovations such as vegetable-based meat substitutes show a potential to disrupt the meat industry and also bring benefits to consumers and possibly increase the sustainability of the food system.

## 5.3.2    Technology and Environment

We address technology and environment together because the environmental dimension of industrial agricultural technology is inextricably linked to the technologies that are being used. Whereas closing yield gaps remains an important priority in the developing world to take full advantage of existing technology, in the industrialized world, yield gaps are small, and most productivity gains are coming from biological, chemical, mechanical, and digital innovations. But, many elements of these technologies are associated with adverse environmental impacts, as per the indicators identified in Tables 3.1, 3.2, 3.3, and 5.4. In the aggregate, the OECD reported some positive elements in the environmental trends for agriculture from 1990 to 2010 (OECD 2013). For example, aggregate use of agricultural pesticides appears to have declined. But, as we emphasized in Chap. 3, the highly diverse and heterogeneous character of agriculture means that aggregate data can be misleading. Indeed, there is evidence of many serious and worsening problems when we look into specific systems and regions. Here, we focus on examples of key elements of agricultural technology and its environmental impacts in the industrialized countries.

## 5.3.3    Biotechnology

We noted in Sect. 5.2 that an increasingly important element of agricultural technology is the use of modern biotechnology (recombinant DNA methods) to accelerate and increase precision of genetic modification and allow the creation of transgenic species (commonly called GM crops or animals). Since the commercial use of agricultural biotechnology began in the 1990s, there has been rapid growth in the use of GM crops (Table 5.5). GM crops are now grown in more than twenty countries and

**Table 5.5** Transgenic crop area for major producing regions, OECD and World (1000 ha)

| | 1996 | 2001 | 2006 | 2011 | 2018 | Share of Arable Crop Land (%) 2008–2011 |
|---|---|---|---|---|---|---|
| USA | 1500 | 35,700 | 54,600 | 69,000 | 75,000 | 39 |
| OECD | 1600 | 39,100 | 61,121 | 80,400 | NA | 18 |
| Brazil | – | – | 11,500 | 30,300 | 51,300 | 30.4 |
| Argentina | 100 | 11,800 | 18,000 | 23,700 | 23,900 | 65.9 |
| India | – | – | 3800 | 10,600 | 11,600 | 5 |
| China | – | 1500 | 3500 | 3900 | 2900 | 3 |
| World | 1700 | 52,600 | 102,201 | 160,000 | 191,700 | 11.4 |

Source: OECD (2013) and ISAAA (2018)

represent about 10% of acres planted worldwide. The United States has the largest total area of GM crops, followed by Argentina, Brazil, and Canada. Other countries such as China and India have increased their use in the past decade but still remain small, with minimal area planted in Europe, Africa, and other regions. The main crops are soybeans, maize, and canola, due to the widespread use of herbicide-tolerant varieties of these crops, with cotton also important for resistance to an important insect pest (the boll weevil). However, it is important to recognize that these figures on area planted may not represent the importance of GM varieties for other crops such as vegetables and fruits, which do not occupy large areas but are of high value and important for nutrition. New varieties are being developed and rapidly adopted, such as a drought-tolerant variety of maize.

Both environmental and related health benefits are attributed to some GM crop and livestock technologies. One example is the use of 'Bt cotton' which is resistant to the boll weevil, allowing farmers around the world, most notably in China and India, to produce cotton with much less insecticide. But on the other hand, the development of soybean, maize, and canola varieties tolerant of the herbicide glyphosate (commercially marketed as Roundup) has led to widespread use of this herbicide, which is now being associated with human health problems, as well as with the development of 'super weeds' that have evolved resistance to the herbicide. Other environmental benefits include the development of better biofuel crops that can substitute for fossil fuels and new drought-tolerant maize varieties that could help farmers adapt to climate change.

### 5.3.4   Nutrients and Water Quality

High rates of the use of nitrogen and phosphorous fertilizer, as well as nutrients contained in animal waste, continue to be a major environmental problem in many parts of the industrialized world. Human health is impacted by contamination of drinking water by nitrates. A major environmental impact is eutrophication of surface water which can harm aquatic life, in some cases leading to hypoxia or 'dead zones' in the Gulf of Mexico and nearly 500 other areas globally (World Resources Institute 2019). Another problem is the occurrence of harmful algal blooms associated with water polluted by phosphorous, for example, in Lake Erie in the United States. Disposal of animal waste also poses serious risks to surface and groundwater quality throughout the industrial world, either from waste storage in ponds or injection into soil (Box 5.2).

---

**Box 5.2   Groundwater Pollution from Animal Waste Disposal in Germany**

Worldwide, animal husbandry produces large amounts of manure that is rich in minerals (phosphates, nitrogen, potassium), as well as proteins, fatty acids, and other important organic compounds. Manure can be used as a fertilizer or energy source, or to make new products, including bio-based ones. However, when industrial methods of intensive, confined livestock production are adopted, more manure may be produced than can be used or disposed of without impacting the environment. This problem has been growing and becoming acute in many parts of Europe. According the Eurostat (2019), nationally averaged nitrate concentrations are all well below the Nitrates Directive and Drinking Water Directive limit of 50 milligrams per liter, but these aggregated data mask the variation found at groundwater monitoring stations; for example, approximately 13% of the stations across Europe were found to exceed the 50 milligram per liter limit in 2009. This has been a growing problem in a number of countries, including Germany, which regulates manure disposal but not well enough to prevent increasing water pollution. For a number of years, the European Union Commission has warned Germany about illegally high levels of nitrate in groundwater that exceed the standard established by the EU Nitrates Directive and pose risks to drinking water. Much of the pollution is attributed to the application of liquid manure to crop fields (European Commission 2018). Other uses of animal waste are to produce methane biogas using anaerobic digestion. However, the economics of biogas versus crop fertilization remains a challenge for farmers, especially those operating at smaller scales (Lauer et al. 2018).

## 5.3.5    Pesticides and Other Chemicals

OECD data show that pesticide and other chemical use in agriculture has not increased in the aggregate over recent decades, but their use remains high in many production systems. Long-standing environmental and health risks associated with pesticides and other chemicals, such as antibiotics used in animal production, continue to be a major problem in industrial systems. Recognition of these problems led to widespread regulation of these chemicals in most industrialized countries, beginning in the 1970s. Major issues include: risks to nontarget insects, notably honey bees; risks to vulnerable birds, fish, and other wildlife; human health risks to farm workers; human health risks associated with pesticide residues in food and drinking water; antibiotic resistance in livestock; and the use of other additives such as the use of recombinant growth hormones in dairy cattle.

## 5.3.6    Greenhouse Gas Emissions

Production agriculture is estimated to be an important emitter of greenhouse gases in both developing and industrial countries. Production agriculture is estimated to contribute about 10–15% of global emissions, whereas the entire food system may contribute as much as one-third of emissions. Agriculture is estimated to be the largest source of non-$CO_2$ greenhouse gas emissions, due to nitrogen fertilizer use causing nitrous oxide emissions from crop fields and due to methane emissions from animals and animal waste. In temperate countries such as the United States and in Western Europe, emissions associated with fossil fuel use, fertilizer use, and livestock production are increasing but are being offset to some degree by improved soil management and reforestation. However, deforestation in some countries, most notably in Brazil, some parts of Africa such as the Congo, and in Indonesia, is a major source of agricultural emissions. Deforestation for both crop and livestock production in the Amazon region increased substantially in 2018–2019 in response to a number of economic and political factors affecting global production and trade (Box 5.3).

**Box 5.3   Trade, Deforestation, and the Challenge of Sustainable Development in the Amazon**

A dramatic increase in deforestation and wildfires in the Amazon region in 2019 illustrates the complex challenge of sustainable agricultural development in a globalized world, thus threatening the Amazon's biodiversity as well as contributing to greenhouse gas emissions. The fires appear to be driven in part by an increase in soybean production and exports from Brazil to China in response to the 'trade war' between China and the United States and the growing demand for pasture land in Brazil for beef production (Jezequel 2019).

Brazil is one of the world's largest exporters of beef, providing about 20% of the global beef exports and is expected to continue to increase that share over the next decade. Brazil also has one of the world's largest numbers of cattle, over 200 million, fed largely with pasture. Since the 1990s, Brazil's cattle herd increased by over 50%. This growth is supported by subsidies for improved pastures and crossbreeding and growing global demand from countries including China, which accounts for 44% of Brazil's exports in 2018 (Economic Research Service 2019b).

In the past three decades, the world's total food production has doubled, while food exports have increased even more rapidly as the food system became increasingly globalized. From the late 1990s until 2018, the United States provided about 40% of China's soybean imports, then due to tariffs imposed between the two countries, that share fell to 10% in the period from October 2018 to June 2019. Historically, the share of Chinese soybean imports from Brazil increased from about 20% in 2000 to about 50% in 2018, then in 2018–2019 alone it jumped from 50% to 70% (Economic Research Service 2019a). Thus, it is clear that about half of the decline in US soybean exports to China due to the trade conflict was replaced with imports from Brazil.

This is happening despite Brazil's efforts to reduce deforestation over the past decade. At the 2009 Copenhagen Climate Conference, Brazil pledged to reduce emissions by over a third by 2020, through a set of land use regulations intended to reduce deforestation and technology improvements that were intended to increase efficiency of livestock production. However, the Bolsonaro government that came to power in 2018 appears to have relaxed enforcement of environmental regulations and encouraged agricultural expansion, contributing to the impact of the global economic forces encouraging deforestation for both soy and beef production.

## 5.3.7   Social Dimensions

Health and safety of farm workers remains an important issue in industrial agriculture, which remains one of the most dangerous industries for workers. Risks include machinery operation and exposure to toxic chemicals.

Mental health also is a concern for farmers and farm family members in developing countries as well as industrialized countries, due in part to isolation in sparsely populated areas and due to financial stresses associated with indebtedness and economic uncertainty. These stresses may be contributing, in part, to the out-migration of younger people and the aging of the farm household populations in most industrialized countries. For example, the average age of farm operators in the United States is about sixty and more than one-third of farmers in Europe are over sixty-five.

Population decline is another challenge for agricultural and rural areas of industrialized countries. A declining population results in a growing mismatch between the supply and demand of services, creating difficulties for both the public and private sectors. As a result of weak local markets, services become underutilized, poorly maintained, and often become unviable and have to be withdrawn. Local living conditions and quality of life deteriorate, unemployment rises, and skilled labor becomes scarce, causing the emergence of abandonment and obsolescence. This further erodes the attractiveness of a region and the rotation of a downward spiral of demographic decline through falling fertility rates and aging of the remaining population. For example, in 2015, the majority of European regions with high shares of elderly populations, and with a corresponding high old-age dependency ratio, were rural. In the United States, 24% of all US counties are losing population. More than 46% of remote rural counties are depopulating compared to 24% of the adjacent nonmetropolitan counties and just 6% of metropolitan counties (Johnson and Lichter 2019).

These social challenges associated with population aging and decline can also be related to agricultural technology and policy. The consolidation of farms into larger units that utilize labor-saving mechanical and biological technologies reduces employment opportunities in rural areas. Conservation policies, such as the Conservation Reserve Program in the United States, that pay land owners to convert crop land to a conserving use such as permanent grass or trees, also reduce the demand for labor and for agribusiness services that can adversely impact rural communities and contribute to demographic and economic decline. Studies in the United States do not show a clear reduction in overall employment (Sullivan et al. 2004), due to offsetting increases in non-agricultural activities such as hunting and recreation. But these analyses show there can be tradeoffs between agriculture and other sectors in rural areas associated with policies designed to reduce adverse environmental impacts of agriculture. One way to reduce these tradeoffs is by 'working lands' policies that allow land to remain in production while also avoiding environmental externalities. For example, in the case of the Conservation Reserve, rules do not allow land in the program to be converted to less-eroding but productive uses such as permanent hay or pasture, even though these uses could substantially reduce soil erosion.

---

### Box 5.4   Improving the Sustainability of Large-Scale Dryland Wheat Production

The dryland wheat production system introduced in Box 3.1 faces sustainability challenges in several dimensions. Soil erosion on steeply sloped hillsides reduces crop productivity and increases water contamination and releases soil carbon and nitrous oxide, two key greenhouse gases, into the atmosphere. The crop-fallow rotation in these systems is used to restore soil moisture during the fallow period, resulting in higher yields when a wheat crop is grown. However, the fallow rotation leaves fields with bare soil every other year, contributes to soil carbon loss and also is vulnerable to wind erosion of soil that pollutes air, and requires tillage or the use of herbicides to prevent excessive weed populations in fallowed fields. Large diesel tractors emit greenhouse gases, as does transportation of crops to distant national and international markets.

There are a number of options for improving the environmental performance of this system. Soil erosion can be substantially reduced by using 'no-till' management that involves leaving crop stubble after harvest, and during the fallow period, and then using special planters called 'seed drills' that can plant a new crop in a field with stubble. Use of precision 'variable rate' nitrogen application technology can reduce fertilizer use and the water pollution and greenhouse gas emissions from the system (see Box 5.1). New digital technologies such as 'plant wearable' sensors have the potential to further improve the performance of precision input application systems. Another option is to eliminate the fallow period by continuously cropping wheat or rotating wheat with another crop. In areas with low rainfall, continuous wheat is economically inferior to the fallow rotation, but use of another crop such as a legume crop (e.g., peas or lentils) can add nitrogen to the soil and reduce erosion, but will also reduce wheat yield if the soil moisture is reduced by the legume crop. Another option is to plant a crop such as *Camelina sativa*, an oilseed crop that can be used to produce biodiesel fuel for jet engines. Elimination of fallow may reduce herbicide use, can increase the amount of soil carbon, and can reduce nitrogen fertilization of wheat, thus reducing the net greenhouse gas emissions from the system.

## 5.4    Conclusion

We conclude this chapter by discussing some of the ways that industrial agricultural systems can move in more sustainable directions. As in Chap. 4, here we focus primarily on the technological options at the system level. In Chap. 6, we extend the discussion to consider how the agricultural systems and sectors more generally can be moved toward sustainable development pathways.

We are all aware that many views about what constitutes the appropriate goals for agricultural sustainability—some people advocate for 'organic' production practices, for example, others argue for 'conservation' or 'ecological' or 'regenerative' agriculture. There is compelling evidence that, in many respects, these types do perform better in some dimensions of sustainability, primarily environmental. Moreover, it seems fair to say that many advocates of these alternative systems would argue that industrial systems, by their very nature, cannot be sustainable—that only small-scale, diversified systems can be truly sustainable as they would have it be defined. But whether or not small-scale diversified systems or larger-scale more specialized industrial systems can be 'sustainable' in all three dimensions—economic, environmental, and social—is the central value judgment in the sustainability debate that cannot be answered by science.

As our discussion of tradeoffs in Chap. 3 illustrated, the economic dimension of sustainability plays a critical role in the debate over how to improve agricultural system sustainability. As we discussed in Sect. 3.4 of Chap. 3, there is an underlying economic explanation for the existence of tradeoffs across the three dimensions of sustainability. This tension between economic performance of agricultural systems and the environmental and social dimensions—at the farm level and at the larger landscape or population scales—is at the heart of the issue and critical to finding solutions. As we noted in Sect. 5.2, the superior economic performance of larger, more specialized farms continues to drive the changes that have occurred since the Green Revolution in the latter half of the twentieth century, and that we continue to see in most major agricultural systems around the world. A major shortcoming of small-scale, diversified systems is their inferior economic performance at the larger scales needed to be competitive with more specialized systems. The reason why many current industrial systems have not moved in directions that perform better in environmental and social dimensions is that farm households would have to bear a cost to do so. As we discuss further in Chap. 6, until now, research and development has been driven by economic considerations, so there have been few mechanisms in place to move technologies at the farm level, and the broader food system, in directions that exploit possible synergies.

The good news is that this situation is beginning to change, and there are indeed existing as well as emerging technologies that can contribute to more sustainable development pathways for industrial agricultural systems, such as the dryland wheat system (Boxes 3.1, 5.1, and 5.4). In Chap. 6 we discuss further this example and the tradeoffs and synergies possible. Similar technological options are available for other major grain-based systems, such as the maize and soybean systems that are increasingly important to the industrial food system for human foods as well as for animal feed and biofuels.

Livestock systems have undergone major changes in the past several decades and also pose major sustainability challenges that can be met with new technologies (HLPE 2016). In countries like the United States, livestock systems have already undergone major changes toward industrial practices, including the use of confinement and drugs such as antibiotics to manage animal health, and these kinds of systems are moving rapidly into transitional countries such as China. In other regions, such as Europe, there is a mix of these types of systems as well as less intensive systems. In addition to the animal welfare concerns associated with confinement and other intensive practices, major sustainability challenges include waste management, greenhouse gas emissions, and use of antibiotics and other drugs. New technologies such as sensors for livestock can monitor health and the reproductive cycle and improve feeding efficiency. Sensors that provide animal location can reduce the cost of grazing systems and improve the utilization of pastures.

# References

Basso, Bruno, and John Antle. 2020. Digital Agronomy to Design and Scale Sustainable Agricultural Systems. *Nature Sustainability*. In press.

Bechdol, Elizabeth, Allen Gray, and Brent Gloy. 2010. Forces Affecting Change in Crop Production Agriculture. *Choices: The Magazine of Food, Farm, and Resource Issues* 25: 11–16.

Beckman, Jayson, John Dyck, and Kari E.R. Heerman. 2017. *The Global Landscape of Agricultural Trade, 1995–2014*. EIB-181. U.S. Department of Agriculture, Economic Research Service.

Bokusheva, R., and S. Kimura. 2016. Cross-Country Comparison of Farm Size Distribution. OECD Food, Agriculture and Fisheries Papers, No. 94. OECD Publishing, Paris. https://doi.org/10.1787/5jlv81sclr35-en.

Economic Research Service, USDA. 2019a. Brazil Once Again Becomes the World's Larges Beef Exporter. https://www.ers.usda.gov/amber-waves/2019/july/brazil-once-again-becomes-the-world-s-largest-beef-exporter/.

———. 2019b. Share of Chinese Soybean Imports Supplied by Brazil has Increased; U.S. Share has Decreased. https://www.ers.usda.gov/data-products/chart-gallery/gallery/chart-detail/?chartId=93777.

European Commission. 2018. Report from the Commission to the Council and the European Parliament on the Implementation of Council Directive 91/676/EEC Concerning the Protection of Waters Against Pollution Caused by Nitrates from Agricultural Sources based on Member State Reports for the Period 2012–2015. https://eur-lex.europa.eu/legal-content/en/TXT/?uri=CELEXpercent3A52018DC0257.

Eurostat. 2019. Statistics Explained: Agri-environmental Indicators—Nitrate Pollution of Water. https://ec.europa.eu/eurostat/statistics-explained/pdfs-cache/16825.pdf.

FABLE. 2019. Pathways to Sustainable Land-Use and Food Systems. 2019 Report of the FABLE Consortium. International Institute for Applied Systems Analysis (IIASA) and Sustainable Development Solutions Network (SDSN), Laxenburg and Paris.

FAO and OECD. 2019. *Background Notes on Sustainable, Productive and Resilient Agro-Food Systems: Value Chains, Human Capital, and the 2030 Agenda*. Rome. http://www.oecd.org/publications/background-notes-on-sustainable-productive-and-resilient-agro-food-systems-dca82200-en.htm.

Freedman, D. 2013. The Truth about Genetically Modified Food. *Scientific American*, September 1. https://www.scientificamerican.com/article/the-truth-about-genetically-modified-food/.

HLPE. 2016. Sustainable Agricultural Development for Food Security and Nutrition: What Roles for Livestock? A report by the High Level Panel of Experts on Food Security and Nutrition of the Committee on World Food Security, Rome.

ISAAA. 2018. Global Status of Commercialized Biotech/GM Crops in 2018: Biotech Crops Continue to Help Meet the Challenges of Increased Population and Climate Change. ISAAA Brief No. 54. ISAAA, Ithaca, NY.

James, Clive. 2013. Global Status of Commercialized Biotech/GM Crops: 2013. ISAAA Brief No. 46. ISAAA, Ithaca, NY.

Jezequel, M. 2019. Global Appetite for Beef, Soy Fuels Amazon Fires. phys.org/news/2019-08-global-appetite-beef-soy-fuels.html.

Johnson, Kenneth M., and Daniel T. Lichter. 2019. Rural Depopulation: Growth and Decline Processes over the Past Century. *Rural Sociology* 84 (1): 3–27. https://doi.org/10.1111/ruso.12266.

Lauer, M., J.K. Hansen, P. Lamers, and D. Thrän. 2018. Making Money from Waste: The Economic Viability of Producing Biogas and Biomethane in the Idaho Dairy Industry. *Applied Energy* 222: 621–636.

MacDonald, James M., Penni Korb, and Robert A. Hoppe. 2013. *Farm Size and the Organization of U.S. Crop Farming*. ERR-152. U.S. Department of Agriculture, Economic Research Service.

MacDonald, James M., and William D. McBride. 2009. *The Transformation of U.S. Livestock Agriculture: Scale, Efficiency and Risks*. EIB-43. U.S. Department of Agriculture, Economic Research Service, January.

OECD. 2013. *OECD Compendium of Agri-environmental Indicators*. Paris: OECD Publishing. https://doi.org/10.1787/9789264186217-en.

Porter, J.R., L. Xie, A.J. Challinor, K. Cochrane, S.M. Howden, M.M. Iqbal, D.B. Lobell, and M.I. Travasso. 2014. Food Security and Food Production Systems. In *Climate Change 2014: Impacts, Adaptation, and Vulnerability. Part A: Global and Sectoral Aspects. Contribution of Working Group II to the Fifth Assessment Report of the Intergovernmental Panel on Climate Change*, ed. C.B. Field, V.R. Barros, D.J. Dokken, K.J. Mach, M.D. Mastrandrea, T.E. Bilir, M. Chatterjee, K.L. Ebi, Y.O. Estrada, R.C. Genova, B. Girma, E.S. Kissel, A.N. Levy, S. MacCracken, P.R. Mastrandrea, and L.L. White, 485–533. Cambridge, UK and New York, NY: Cambridge University Press.

Salin, Delmy. 2018. Soybean Transportation Guide: Brazil 2017. U.S.421 Department of Agriculture, Agricultural Marketing Service. https://doi.org/10.9752/TS048.09-2018.

Sullivan, P., D. Hellerstein, L. Hansen, R. Johansson, S. Koenig, R. Lubowski, W. McBride, et al. 2004. The Conservation Reserve Program: Economic Implications for Rural America. ERS, Agricultural Economic Report 834, October. U.S. Department of Agriculture, Washington, DC.

Thematic Group on Sustainable Agriculture and Food Systems. 2013. Solutions for Sustainable Agriculture and Food Systems. http://www.dun-eumena.com/reagri/upload/files/130919-TG07-Agriculture-Report-WEB.pdf.

World Health Organization. 2014. Frequently Asked Questions on Genetically Modified Foods. https://www.who.int/foodsafety/areas_work/food-technology/faq-genetically-modified-food/en/.

World Resources Institute. 2019. Interactive Map of Eutrophication and Hypoxia. https://www.wri.org/resource/interactive-map-eutrophication-hypoxia.

# 6

# Pathways to Sustainable Agricultural Development

## 6.1 Introduction

In this chapter, our aim is to discuss ways that we can, individually and collectively, move agricultural development in more sustainable directions. As we have discussed in the preceding chapters, the complexity of agricultural systems and their linkages to the broader economy and society means that there are inevitable tradeoffs among economic, environmental, and social outcomes dictated by physical, biological, economic, social, and political realities. Recognizing that there will be tradeoffs among goals such as the SDGs is not an argument for pessimism, but rather a call for a truer understanding of the complex reality in which we live, and a commitment to meet the challenge before us. Despite the daunting challenges humanity faces, we can take succor from the formidable progress that has been made in many dimensions of sustainable development over the past century. Fifty years ago, many people justifiably doubted the ability of the world to feed itself and prevent a Malthusian future of recurring mass famines. That dire future was avoided through applications of science and technology as well as institutional and policy innovations that created a remarkable global food system. Yet,

© The Author(s) 2020
J. M. Antle, S. Ray, *Sustainable Agricultural Development*, Palgrave Studies in Agricultural Economics and Food Policy, https://doi.org/10.1007/978-3-030-34599-0_6

as we have discussed in the previous chapters, these very successes in agricultural development, and in broader economic development, have created huge challenges in environmental and social dimensions that were largely unanticipated when the Green Revolution technologies of the twentieth century were making possible rapid growth in global food production, consumption, and trade, creating the basis for the globalized food system of today. The good news of this chapter is that we now have many tools we can use to design and build more sustainable agricultural systems. But we will conclude that the challenge of using those tools effectively to move the diverse agricultural systems of the world in sustainable directions remains daunting.

The complexity of agricultural systems—what we described in Chap. 3 as their diversity and heterogeneity—was elaborated in Chaps. 4 and 5 where we discussed the characteristics of agricultural systems in developing, transitional, and industrialized countries. This complexity gives rise to various synergies and tradeoffs as we attempt to improve the performance of these systems, and it means simple, one-size-fits-all solutions to sustainable development rarely, if ever, exist. The key implication for the design of sustainable development pathways is that solutions must be tailored to fit the problem—to the specific system, the location, and the geographic and temporal scales. In this chapter, we elaborate the two essential steps in moving agriculture as a whole (Chap. 2) or individual agricultural systems (Chap. 3) toward more sustainable pathways: first, we must *design* or *envision* alternative pathways that move the sector or system toward sustainable development goals; second, we must *implement* actions that will move the sector or a system toward those goals.

Today, many organizations are using participatory, science-based approaches to evaluate current 'business as usual' or BAU pathways and alternative pathways that could improve the sustainability of agricultural systems. While there are many variations on these approaches, most now build on the idea that sustainability encompasses the positive (data and science-based) and normative (values and goals) elements. The design step involves two main components. In Sect. 6.2, we discuss the normative component that involves scientists and stakeholders working together to identify plausible future 'states of the world' and goals that could be achieved along alternative, more sustainable development pathways. In Sect. 6.3, we discuss the second element of the design step: the tools and methods—the data and science—needed to understand and project the performance of agricultural systems along alternative development pathways envisioned in the first step. Finally, in Sect. 6.4, we discuss the final, critical step: implementation—how to achieve the technologies and policies needed to realize the goals envisioned for sustainable development.

## 6.2    Participatory Pathway Design

By its very definition, sustainable development must meet the needs of all people, and so the design of sustainable development pathways must involve broad participation. Thus, at the global scale of the sustainable development goals, the community of nations participates through processes overseen by the United Nations. Individual countries and regions within countries, in turn, have their own political and social processes, and meet this standard of broad participation to varying degrees. In this section, we describe the approaches that civil society and the science community are using to design pathways at the global, national, and local (agricultural system–level) scales.

The pathway concept now being used in science derives originally from the military's use of 'scenarios' in war planning. The term scenario is now widely used in scientific computer modeling of complex systems, as we discuss further in Sect. 6.3. A major impetus for scenario methods has been the development of earth system science and global climate modeling, much of it aiming to evaluate global climate changes occurring because of increasing concentrations of anthropogenic greenhouse gases. In climate science, greenhouse gas emissions scenarios are developed based on projections of economic development, and these emissions scenarios are used to drive climate model simulations. As we shall discuss later, a similar methodology is used to simulate economic development at global and national scales, as well as agricultural system performance, in response to climate change as well as other types of changes such as policy or technology.

**Table 6.1** Proposed global targets for sustainable land use and food systems

| Area | Global Target |
|---|---|
| Food security | Zero hunger |
| | Average daily energy intake per capita higher than the minimum requirement in all countries by 2030 |
| | Low dietary disease risk |
| | Diet composition to achieve premature diet-related mortality below 5% |
| Greenhouse gas emissions | Greenhouse gas emissions from crops and livestock compatible with keeping the rise in average global temperatures to well below 1.5 °C |
| | Below 4 GtCO$_2$e per year by 2050 |
| | Greenhouse gas emissions and removals from land use, land-use change, and forestry (LULUCF) compatible with keeping the rise in average global temperatures to below 1.5 °C |
| | Negative global greenhouse gas emissions from LULUCF by 2050 |
| Biodiversity and ecosystem services | A minimum share of earth's terrestrial land supports biodiversity conservation |
| | At least 50% of global terrestrial area by 2050 |
| | A minimum share of earth's terrestrial land is within protected areas |
| | At least 17% of global terrestrial area intact by 2030 |
| Forests | Zero net deforestation |
| | Forest gain should at least compensate for the forest loss at the global level by 2030 |
| Freshwater | Water use in agriculture within the limits of internally renewable water resource, taking account of other human water uses and environmental water flows |
| | Blue water use for irrigation <2453 km$^3$ per year (670–4044 km$^3$ per year) given future possible range (61–90%) in other competing water uses |
| Nitrogen | Nitrogen release from agriculture within environmental limits |
| | N use <69 Tg N per year total industrial and agricultural biological fixation (52–113 Tg N per year) and N loss from agricultural land <90 Tg P per year from freshwater systems into ocean by 2050 |
| Phosphorous | Phosphorous release from agriculture within environmental limits |
| | P use <16 Tg P per year from fertilizers to erodible soils (6.2–17 Tg P per year), P loss from agricultural soils, human excretion <8.69 Tg P per year flow from freshwater systems into ocean by 2050 |

Source: FABLE (2019)

Recent advances in climate and sustainability research have refined these pathway methods for analysis of agricultural systems at global, national, and subnational scales. An example at the global and national scales are the global targets developed by the Food, Agriculture, Biodiversity, Land use, and Energy (FABLE) Consortium (FABLE 2019) for land use and food system sustainable development (Table 6.1). These goals are informed by various other goal-setting activities including the SDGs, climate change research programs, and a consortium of private and public interest groups and scientists from eighteen countries. As we can see from Table 6.1, these targets are closely aligned with the indicators we identified for agricultural systems sustainability (Tables 3.1, 3.2, 3.3) and used by organizations such as the Organisation for Economic Co-operation and Development (OECD) and European Union (Table 5.4). Box 6.1 further elaborates the process the FABLE Consortium uses.

---

**Box 6.1   The FABLE Consortium: Designing Global and National Pathways for Sustainable Land Use and Food Systems**

The FABLE (Food, Agriculture, Biodiversity, Land, and Energy) Consortium is part of the Sustainable Development Solutions Network. The FABLE Consortium was created as a global network of researchers from eighteen developed and developing countries to build tools and analyses for integrated land- and water-use planning. The Consortium supports country teams to develop rigorous, transparent pathways toward sustainable land use and food systems that demonstrate the feasibility of rapid progress and help raise the level of ambition toward the SDGs and the objectives of the Paris Agreement (to limit global warming to 1.5 °C) reached at the twenty-first Conference of the Parties of the United Nations Framework Convention on Climate Change held in December 2016. To this end, the Consortium pursues three broad sets of activities: capacity development and sharing of best practice for data management and modeling of the three pillars; development of mid-century national pathways that can collectively achieve the jointly agreed global targets and have consistent trade assumptions; and analysis of national policy options and support to national and international policy processes. FABLE has developed a modeling tool that can be used by national teams to evaluate the implications of the global targets in Table 6.1 for national land use and food systems, and the contributions of national strategies to the global goals.

Another example of pathway design is AgMIP's framework for Coordinated Global and Regional Assessment (Fig. 6.1). The core pathways include business as usual and climate change based on projected emissions and socio-economic pathways (similar to those discussed in Box 4.1), and sustainable development pathways. These core pathways can incorporate various other elements, in this case grouped into four areas: greenhouse gas mitigation; climate adaptation; food policy; and food security.

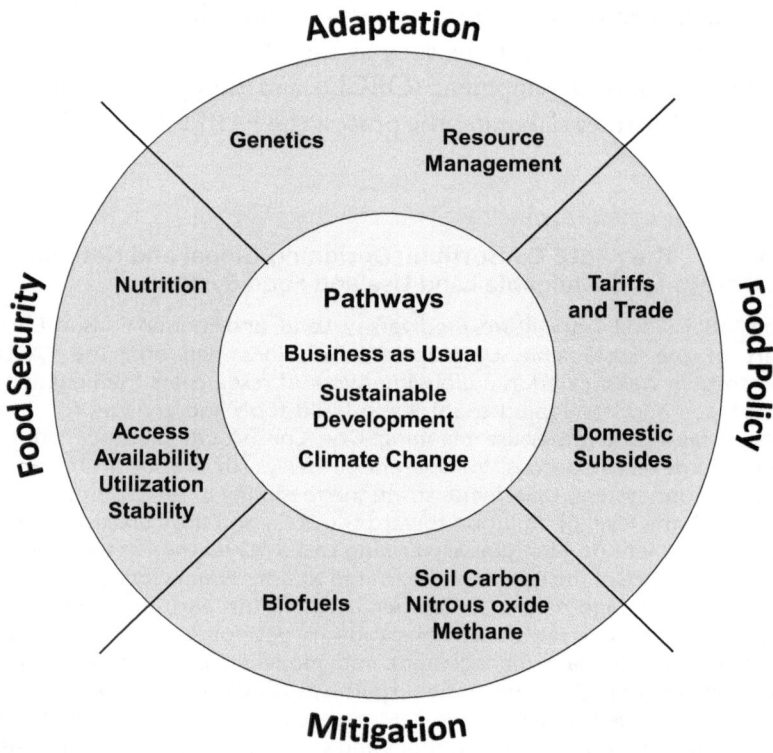

**Fig. 6.1** AgMIP's pathways for Coordinated Global and Regional Assessment. (Source: Based on Rosenzweig et al. [2016])

At the agricultural system level, participatory approaches are now being widely used to develop and assess agricultural technologies and development pathways. The example we point to here is 'Representative Agricultural Pathways' or RAPs (Valdivia et al. 2015) (Box 6.2). RAPs are developed with teams made up of scientists and stakeholders to support the tradeoff analysis process (Fig. 3.8). The development of RAPs begins with a narrative description of a future state of the world, followed by identification of key features of future systems in bio-physical, technology, economic, and social dimensions, as well as identification of key sustainability indicators to be used for evaluation of pathways (as in Table 3.1). Goals can be identified for key indicators, either by the stakeholder group or by linking the pathway concepts to goals established by governmental processes. For example, Box 6.6 presents analysis of development pathways for Kenya's crop-livestock systems, based on goals established by the Kenyan government to reduce poverty and improve environmental sustainability.

---

**Box 6.2   Representative Agricultural Pathways for Kenya's Vision 2030 Plan**

The Kenyan government's economic development plan known as Vision 2030 aims to achieve sustainable development goals including lower poverty and higher agricultural productivity. This plan was developed with broad participation among governmental and nongovernmental organizations. To assess the implications of this plan for the crop-livestock systems in Machakos, Kenya, two RAPs were developed to define the likely changes in key drivers of the system associated with the Vision 2030 policies. Figure 6.2 provides a qualitative picture of how these policies were translated into qualitative changes in key technological and economic factors. The research team then translated these qualitative changes into plausible quantitative changes for simulation modeling. The results of this analysis are presented in Box 6.6.

*(continued)*

**Box 6.2  (continued)**

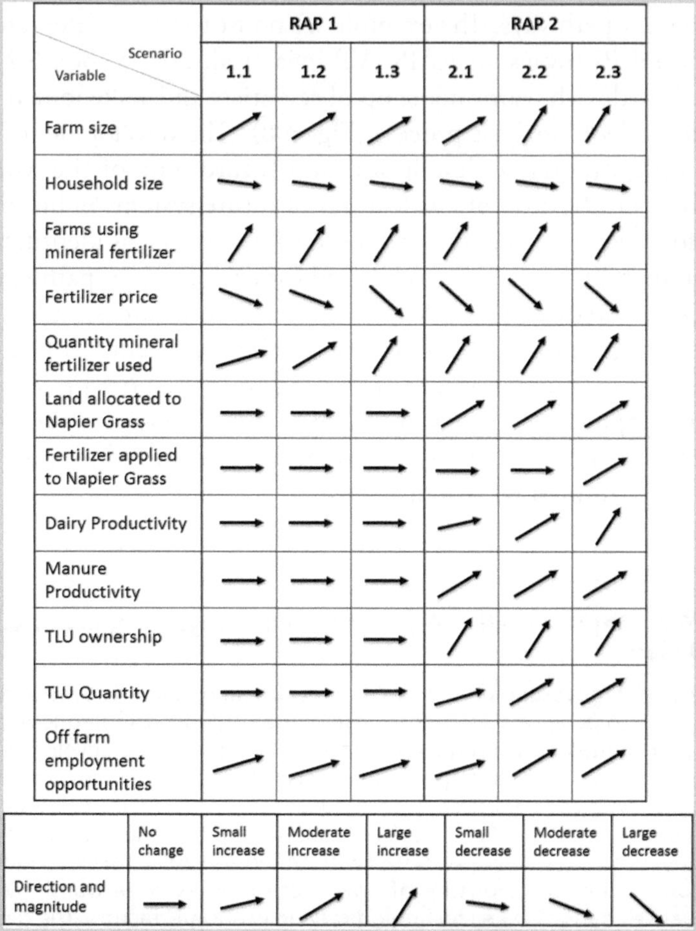

| Scenario / Variable | RAP 1 | | | RAP 2 | | |
|---|---|---|---|---|---|---|
| | 1.1 | 1.2 | 1.3 | 2.1 | 2.2 | 2.3 |
| Farm size | moderate increase | moderate increase | moderate increase | moderate increase | large increase | large increase |
| Household size | no change | no change | no change | no change | no change | no change |
| Farms using mineral fertilizer | large increase | large increase | large increase | large increase | large increase | large increase |
| Fertilizer price | small decrease | small decrease | moderate decrease | moderate decrease | moderate decrease | moderate decrease |
| Quantity mineral fertilizer used | small increase | moderate increase | large increase | moderate increase | moderate increase | large increase |
| Land allocated to Napier Grass | no change | no change | no change | moderate increase | moderate increase | moderate increase |
| Fertilizer applied to Napier Grass | no change | no change | no change | no change | no change | moderate increase |
| Dairy Productivity | no change | no change | no change | small increase | moderate increase | large increase |
| Manure Productivity | no change | no change | no change | moderate increase | moderate increase | moderate increase |
| TLU ownership | no change | no change | no change | large increase | large increase | large increase |
| TLU Quantity | no change | no change | no change | moderate increase | moderate increase | moderate increase |
| Off farm employment opportunities | moderate increase | small increase | small increase | small increase | moderate increase | moderate increase |

| | No change | Small increase | Moderate increase | Large increase | Small decrease | Moderate decrease | Large decrease |
|---|---|---|---|---|---|---|---|
| Direction and magnitude | → | ↗ (small) | ↗ (moderate) | ↑ (large) | ↘ (small) | ↘ (moderate) | ↘ (large) |

**Fig. 6.2** Representative Agricultural Pathways for Machakos, Kenya, based on the Kenyan Government's Vision 2030 plan. (Source: Valdivia et al. [2017]) TLU = tropical livestock unit

## 6.3    Evaluating Agricultural Development Pathways

The goal of improving the sustainability of agricultural systems and moving the agricultural sector in a more sustainable direction creates a major challenge for the scientific community. In response to this challenge, new methods and new data are being developed. To understand this challenge, we can think about what we mean by the 'scientific method.' To most people, including most scientists, the sine qua non of science is the controlled experiment. Indeed, experimentation has been the basis of most agricultural science. The recognition of the public good value of systematic investigation and dissemination of knowledge in the late nineteenth and early twentieth centuries led to the development of public research and extension systems, as exemplified by the US land grant university experiment station system. This system exploited key breakthroughs in statistical methodology in the early twentieth century, based on the application of new mathematics to formalize the central role of randomization in experimental design and hypothesis testing.

Yet, experimentation as the foundation of the scientific method has its limits. One problem is that, even for those phenomena that can be studied by experimental science, the process remains a slow one, constrained by the rates of plant and animal growth, their reproductive cycles, and the annual climate cycle. The experimental method also is constrained by the large number of factors that can affect crop or animal growth and other characteristics associated with its genetics, environment, and management—an experiment that varies one factor, holding all else constant, may tell little about the outcomes that will be observed in the field. In evaluating agricultural systems performance, the consideration of environmental and social sustainability further increases this 'dimensionality problem' in experimental design. And as we move from the scale of the cell to the plant, the plant to the field and farm, and from the farm to the region and the globe, it is evident that controlled experimentation on actual or prospective systems is not feasible. Indeed, the most important questions about sustainable development and climate change involve the performance of systems that do not yet exist, such as systems that use more climate-resilient crop varieties, or that incorporate new crop or livestock species. Moreover, for sustainability research, we need to evaluate how both current systems and new systems would perform under conditions not yet observed, for example, under a projected future climate as well as evolving economic and social conditions. For such analyses, scientists use computer simulation models designed to represent the major features and behavior of these complex systems. As we have already discussed in Chap. 3, models have been developed for the major components of farm household systems illustrated in Fig. 3.1, and also for regional, national, and global agricultural systems.

What are these models? On the supply side, in both agronomy and economics, the 'production function' is one of the central concepts. Economists have used production function models to quantify agricultural technologies, beginning with early pioneers such as Heady (1961). Agronomic researchers began to develop mechanistic crop growth simulation models beginning in the 1960s, encouraged by improved understanding of the relevant processes of crop growth. Initially used to embody process understanding in quantitative terms, researchers combined these models with experimental data and exploited the increasing capability of digital computers to develop numerical simulation models (Jones et al. 2017). Models that began as crop growth models have evolved into larger and more complex system models with various capabilities and are now used for many research, analytical, and decision support purposes, representing individual crops, multi-crop systems, and crop-livestock systems, at scales ranging from an individual point in space (i.e., an individual plant or experimental field) to global crop yield simulations that use spatially referenced soil and climate data to assess the impacts of climate change. Today, these models are linked to economic models for analysis at the farm, national, and global scales (Antle and Stöckle 2017). A global community of science, the Agricultural Model Inter-comparison and Improvement Project (AgMIP), now supports collaborative research on data, software, and model improvement for agricultural systems models (Rosenzweig et al. 2013, 2018).

Most evaluations of agricultural system sustainability, as well as climate impact assessments, are implemented in a framework which integrates climate, crop, livestock, and economic data and models. Large-scale global assessments are implemented using components linked as illustrated in Fig. 3.7. Future climate simulations are the first component of the assessment framework, and are based on assumed trajectories of greenhouse gas emissions that are consistent with a plausible range of future socio-economic conditions. For the agricultural and economic model components, additional socio-economic pathway elements are needed, such as rates of population growth and rates of technological change. The agricultural and economic models project production, consumption, prices, and related outcomes at various spatial and temporal scales, depending on the type of model and objectives of the analysis.

This type of integrated assessment method is being used at scales ranging from global to regional (i.e., multinational or national), and subnational (i.e., the US Corn Belt). For example, Nelson et al. (2013) present results from ten global models that are being used for climate impact assessments. The US Department of Agriculture's Economic Research Service has developed a regional economic impact assessment model that is linked to a bio-physical model that simulates crop yields and other environmental outcomes for agro-ecological regions of the United States (Johansson et al. 2007). van Ittersum et al. (2008) describe an integrated assessment framework created for the European Union. This framework links field-level bio-physical models with farm-level economic optimization models and a multicountry econometric policy model, and is being used for policy analysis and climate impact assessment. van Wijk et al. (2014) review a large number of studies that combine various types of bio-physical simulation models with economic optimization or simulation models to study the sustainability of agricultural systems. One of these models is the TOA-MD model that simulates populations of farm household for assessment of agricultural system sustainability (Box 3.3).

As we noted above, computer simulation models are used to study agricultural system sustainability because the conventional laboratory or field experiments cannot be used. Instead, scientists use these models to conduct simulation experiments. The goal of a simulation experiment is to examine the effects of changing one or more components of the system, while holding other components fixed. For example, in Sect. 3.6.1, we described the simulation experiments being used to study the impacts of climate change and climate adaptations.

The diversity and heterogeneity of agricultural systems that we highlighted in Chaps. 3, 4 and 5 create many challenges for agricultural system modeling. Most notably, these characteristics mean that large amounts of highly detailed, site- and time-specific data are needed. Thus, a key limitation to the development and use of agricultural systems models is data (Antle et al. 2017). Better data are needed to further improve crop and livestock models in ways that are useful for both on-farm management decision-making and for use in research to develop and test new technologies, and to evaluate their productivity and sustainability. Fortunately, new technologies, such as the use of mobile sensors and other 'big data' technology, are helping to bridge this gap, such as the technologies enabling digital agronomy and precision agriculture (Box 5.1). However, a number of important challenges must be overcome to make these data useful for both farmers and scientists. One issue is how to share individual data while maintaining farmers' privacy and property rights to their data. Another issue is how to translate individual data, typically acquired using various nonstandard, proprietary formats, into a generic format that would be FAIR (findable, accessible, interoperable, and reusable). Ideally, an integrated private-public data infrastructure that meets both private and public needs would make sense, but we are far from such a system today. Private data and related soft and hard infrastructure are being developed by a growing array of management advisory and technology companies. Data generated by individual producers or by private firms selling data or advisory services are not public and thus not findable or accessible, often even by farmers themselves. There are no established data standards being used, and thus data are not interoperable even when findable and accessible.

There are also many limitations of currently produced public data. Some of the currently public data, such as some weather, price, and crop yield data, are open access or available for a fee. However, many of the data related to agricultural production are collected for various government administrative purposes and are not intended to be used for research or for private decision-making, and are not easily accessed or used for private or research purposes. Various efforts are underway to address these data challenges (see Sect. 3.6.3).

## 6.4 Implementation: Individual and Collective Action

The second, critical step toward sustainable agricultural development pathways is implementation of actions that will move agricultural systems in more sustainable directions. In the previous section, we discussed participatory processes to define sustainable development goals and pathways, and methods to evaluate them. But identifying goals and associated indicators is only a first step toward achieving more sustainable development pathways at local, national, and global levels. Even if there were a complete agreement about these goals, we have to recognize that the global community of nations cannot design and enforce policies to attain them because there is no global authority to enforce such policies. In fact, in the post-Second World War era, the global community has made efforts to design and enforce policies, for example, in key economic areas such as trade policy through the creation of the World Trade Organization. And despite much progress toward a rational global trading system, recent experience with the China-US 'trade war' has demonstrated the limits of that organization to enforce its rules on the use of policy instruments such as tariffs. Similar experiences have been seen in environmental and social dimensions as well. Thus, we must recognize that designing and enforcing policies to achieve sustainable development goals is primarily the responsibility of national governments, as has become evident from the experience with climate policy (Box 6.3).

**Box 6.3   Lessons from the Montreal and Kyoto Protocols for Sustainable Development**

Two efforts to address global externalities illustrate the challenges for coordinated global efforts to achieve sustainable development. The Montreal Protocol addressed the problem of a 'hole' in the earth's atmospheric ozone layer caused by emissions of chlorinated flurocarbons (CFCs) and related compounds used as refrigerants and for other industrial processes. Four countries were the principal sources of these chemical emissions, and as a result, it was possible for them to reach an accord known as the Montreal Protocol in 1987 to limit or ban their use. The Montreal Protocol appears to have been successful for several reasons: a small number of countries were causing the problem; the problem presented a clear and present danger; and low-cost solutions (technological alternatives) were available. In contrast, the Kyoto Protocol was an attempt by the global community of nations to address a key dimension of sustainable development: climate change.

The Kyoto Protocol attempted to coerce countries to adopt policies that were perceived as potentially imposing a high cost on industrialized countries, to address a problem with highly uncertain consequences in the distant future. After the Clinton administration pushed for the use of cap-and-trade policies to implement the Kyoto Protocol, George W. Bush reneged on his support for the Protocol after his election in 2000.

A majority of countries that supported the Kyoto Protocol did not meet their emissions reduction targets when the agreement expired in 2012. Many argue that Kyoto's failure was due to deficiencies in the structure of the agreement, such as the attempt to impose mandatory emissions reductions on countries through an international agreement, the exemption of developing countries from required emissions reductions, such as China and India, that were becoming major emitters from reduction requirements, and the failure of the largest greenhouse gas emitter at that time (the United States) to support the agreement. Subsequently, global climate policy efforts focused on a system of nonmandatory 'intended nationally determined commitments' to reduce greenhouse gas emissions (World Bank 2019).

How the direction that development in a national economy, or a major sector of an economy such as agriculture, can be influenced depends on how production and consumption decisions are made in the economy. Since the breakup of the Soviet Union in the early 1990s and the liberalization of the Chinese economy beginning in the 1980s, few economies in the world can be described as 'centrally planned.' Rather, most national economies today can be described as market-based, meaning that a substantial share of the production and consumption decisions in the economy is made by privately owned firms and households. However, almost all governments exert substantial control over important parts of their resources and economies. For example, in the United States, the federal and state governments control about 40% of the land area, with the remainder held by private businesses or individuals, but the energy sector is privately owned and managed. In many countries, such as Brazil and Saudi Arabia, the energy sector has been nationalized and is controlled partly or wholly by the government. In most countries, agricultural land is either privately owned, or if state owned, individuals have 'use rights' but not the right to buy or sell the land. In China and the former Soviet states, a transition from state ownership of virtually all land to an increasing degree of privatization has been ongoing for several decades. Thus, the vast majority of household consumption decisions, and a large share of production decisions in agriculture and most other sectors of most economies, are made by private individuals or businesses.

So as individuals and businesses learn more about the sustainability challenges the world faces, why don't concerned individuals and businesses just 'do the right thing'—can't we achieve a more sustainable development pathway if people take responsibility for the consequences of their consumption and production choices?

We will argue below in Sect. 6.4 that there is, indeed, a role for individual choices to influence the development pathway of an economy and of agriculture. But there are also compelling reasons to doubt that individual actions, by themselves, will be adequate to achieve sustainable development pathways. The first reason is that even if individuals are willing to 'do the right thing,' they may have different knowledge or values, and may focus on some aspects of sustainability and ignore others. Combined with the likelihood of tradeoffs among the dimensions of sustainability, individual actions are not likely to achieve a reasoned balancing of competing objectives.

But there are also other fundamental reasons why individual action is not likely to achieve sustainable development goals in market-based economies. To see why, it is important to understand the way that resources are allocated by firms and households. In market economies, production and consumption decisions are based on market prices that reflect the values of consumers (reflected in market demand) and the costs of producing and selling products (reflected in market supply). Experience shows that when market prices accurately reflect the relative scarcity of resources—that is, they reflect the true cost or value of the goods being produced and consumed—then markets do a remarkably good job of providing people with things they need and want. However, we have also learned that in some cases, market-based economies do *not* allocate resources well—for example, most production activities also create air and water pollution. When this happens, economists say there is *market failure*. Why does this happen?

In many cases, we can trace these problems to *common property resources* like air, water, and genetic diversity that are not 'owned' by anyone and thus individuals do not have *incentives* to manage those resources in ways that collectively make sense. A classic example is the overuse of a common property resource like open-access rangelands. Each person using the range sees the value they obtain from using it to sustain their livestock, but they do not have to bear the cost they impose on their neighbors who also want to use the range. The result is often *overgrazing*, that is, the total number of animals using the rangeland reduces its productivity to the point that the value of the rangeland is far less than if the number of animals using it was reduced. In some cases, these problems can be solved through a sharing arrangement— that is, the users of the range meet and agree to limit their use so as to maintain the productivity of the land. This kind of sharing arrangement can work if there is a low-cost way for the sharing to be enforced. In effect, what the users of the rangeland are doing is equivalent to *assigning property rights*, that is, another solution is to divide the range among those using it, each with their own area so they can control the amount of grazing on each parcel (perhaps by constructing fences). Similar situations exist with other open-access common property resources. Overuse of groundwater and overfishing in the oceans are other classic examples.

As the grazing example illustrates, markets fail to allocate resources well when the actions of one or more individuals (say, in producing a product) impose costs on others, and those costs are not borne by the resource users. Health and environmental damages caused by air and water pollution are other prime examples of such 'external costs.' For example, when farmers use fertilizer that runs off their fields in the US Midwest and contribute to hypoxia in the Gulf of Mexico, the farmers are imposing a cost, in the form of the damages caused by nutrient pollution of water, on others such as the fishing industry in the Gulf. But under current laws and policies in the United States, Midwestern farmers are not responsible for these costs.

Similarly, every individual who drives a gasoline-powered automobile in the city of Los Angeles contributes to air pollution that imposes health costs on the residents of Los Angeles. These automobiles also contribute to greenhouse gas concentrations in the atmosphere that in turn cause climate change and impose costs on farmers in the Midwest in the form of more extreme weather events such as droughts and floods. These floods further aggravate the runoff of fertilizers that pollute the Gulf of Mexico!

So why don't farmers reduce their use of fertilizer voluntarily, if they know they are causing damage to fisheries in the Gulf? What don't people just stop driving gas-guzzling cars in Los Angeles? Farmers use fertilizers because they provide value in the form of increased productivity and income. Thus farmers would have to be willing to bear a cost—in the form of reduced productivity and income—to reduce their use of fertilizer. Likewise, people who drive cars do so because they need to get to their jobs each day in Los Angeles. Forgoing the use of a car in Los Angeles would represent a large cost to most people in forgone earnings, higher costs of shopping and child care, and lost recreational opportunities.

While some farmers may be willing to reduce their fertilizer use, and some people may be willing to ride a bike rather than drive a car, they each can rationally say that by acting alone they would not have a discernible effect on the environment, and thus have little reason to reduce their fertilizer use or their use of a car. Each individual can rationally conclude, in other words, that the problem can only be solved if most farmers reduce their fertilizer use, or if most people drive cars less; and if everyone else would reduce their fertilizer use, or drive less, then each individual can also conclude that he or she does not need to—that is, there is an incentive to 'free ride' on the actions of others. Note also that this free-rider problem can be solved in small groups when it is easy for each individual to observe the actions of others. But in large groups, this 'self-enforcement' does not work well. A similar problem arises between people living now and in the future. For example, consuming fossil fuels today may impose costs on future generations through climate change—but what incentive does each individual today have to reduce their fossil fuel consumption, if billions of people in the world are going to continue using fossil fuels anyway?

We can conclude, therefore, that solutions to many of the challenges we face in achieving a more sustainable development pathway require some type of collective action. There are two ways that such collective action can be implemented. One is through direct coercion by the state, for example, if a farmer's use of fertilizer is polluting water, a government regulation can restrict fertilizer use. This type of direct regulation of production activities—often referred to as 'command and control' regulation—is used by many governments. However, there is a critical problem with command-and-control regulation, namely that it often is a very costly way to manage resources in a market economy. The reason is that the amount of pollution caused by, say, each farmer using fertilizer is likely to be different. Some farmers are close to a river and thus cause much more pollution than farmers far away from a river. Thus, if both farmers are required by a regulation to reduce fertilizer use by the same amount, and thus bear a cost in the form of reduced productivity and income, the regulation is inefficient—most of the benefit of the regulation could be achieved by limiting the fertilizer use of the farmer near the river. Forcing both farmers to reduce fertilizer use achieves the desired goal—reduced pollution—at a higher cost than if the regulation were targeted at the farmers causing the most pollution. Thus, we can see that the heterogeneity of farms, in terms of how much pollution they cause, creates a serious challenge to environmental regulation.

An important alternative to command-and-control regulation is the use of *incentive mechanisms*. Various types of incentive mechanisms have been devised to deal with various types of externality problems. One example of an incentive mechanism is a tax on a polluting activity, for example, a tax on fertilizer. However, we can see that a uniform tax on fertilizer also would not necessarily be an efficient solution to the water pollution problem caused by fertilizers if farms are heterogeneous—that is, if some farms cause more pollution than other farms. The challenge is to find a way to better target the regulations or incentives to the individuals causing the externality. One way to do this is for regulators to have information that allows them to identify the individuals who are causing the pollution. For example, highly accurate spatial data on the location of farm fields, combined with computer simulation models and field observations, can now identify the fields that are contributing to water pollution. With this information, a regulatory policy can impose stronger restrictions on farm fields that are most likely to contribute large amounts of nutrient pollution, or can target incentives toward those farms to encourage them to reduce fertilizer, thus avoiding the imposition of costly restrictions on fields that do not contribute much to nutrient pollution.

This nutrient pollution example also illustrates another important aspect of environmental policy—who pays for pollution? Observe that the farmers causing the most pollution could be offered incentive payments to 'voluntarily' reduce their fertilizer use, rather than having their fertilizer use restricted by regulations or by having to pay a fertilizer tax. These incentive payments would be funded by taxing other citizens. While the same environmental outcome would be achieved by a tax, a regulation, or a subsidy, a very different economic outcome would be achieved: the tax or regulation would impose the cost of pollution on the polluter, whereas, the subsidy would impose the cost on all taxpayers. Which solution is used is a political decision that may reflect the relative political power of interest groups in the society—that is, farmers may be better politically organized than taxpayers, and argue that they have a 'right to farm,' whereas, others in society may claim a 'right to clean water.'

Another mechanism that has been invented by economists is the allocation of tradable 'pollution permits.' For example, if it can be determined what level of emissions into air or water can be tolerated, then permits allowing that much pollution to occur can be allocated to producers, as we discuss in the next sections (Box 6.4).

## 6.4.1  Technological Innovation and Sustainable Development

Chapters 4 and 5 identify some of the technologies that could improve sustainability, in both developing and industrial agricultures. But where do they come from? And if technologies are part of the problem, how can they become part of the solution?

As we have discussed in earlier chapters, technological innovation was the foundation of the twentieth century's Green Revolution that made it possible to feed a rapidly growing global population. But these technologies have turned out to be unsustainable in the ways that we described in Chaps. 4 and 5. Farm households throughout the poorer regions of the world still struggle to be economically sustainable and food secure; large-scale, industrial agriculture is largely dependent on fossil fuel-based energy, just as other sectors of industrial economies are. Thus, it seems logical to argue that technological innovations will also be needed to move the major agricultural systems of the world in more sustainable directions, by addressing their economic, environmental, and social shortcomings.

But what will drive the creation of more sustainable agricultural technologies? One of the long-standing economic puzzles is to understand what causes technological change. Sustainable development adds further complexity to this puzzle—we are asking not only about the overall rate of technological improvement, as evidenced by increases in agricultural productivity (say, crop yield), but also about other aspects of the technology that have not only economic but also environmental and social implications.

The history of technological innovation provides some insight into this question. It remains difficult for historians or economists to explain why the industrial revolution occurred when and where it did. But it is clear that technological innovation is not, as some economic models assume, costless 'manna from heaven.' Rather, most technological innovation comes from purposeful investment in scientific research and its application, with these investments being made by research universities and institutes, government agencies, and private industry.

In addition to the overall level of technological innovation, as evidenced by increases in productivity, it is clear that the types of innovations that are developed and adopted are influenced by economic conditions. The classic study of agricultural innovation by Hayami and Ruttan (1971) compared the Japanese and United States experiences. Hayami and Ruttan showed that in countries like Japan, where land is very scarce and labor is relatively abundant, Japanese agricultural innovation strongly favored biological innovations, such as higher-yielding varieties or rice that responds well to intensive cultivation with fertilizers and human labor on small plots of land. In contrast, in the United States, where land and capital are abundant and labor relatively scarce, innovation has favored biological and mechanical technologies that are land and capital using and labor saving. We use the term *induced innovation* to describe this process of economic conditions influencing technological innovation.

The concept of induced innovation has important implications for sustainable development. As we noted earlier, market economies allocate resources largely based on market prices as signals of resource scarcity and consumer demand. But we also noted that markets that are responding to private benefits and costs fail to account for externalities or other non-market considerations such as equity or concerns about animal welfare. It follows then that innovations induced by price signals that reflect only private benefits and costs will not take these non-market factors into account. This fact then helps us understand why the 'Green Revolution' technologies developed in the twentieth century did not take into consideration, for example, the human health or environmental effects of dangerous pesticides, and why the development of industrial agricultural systems does not today take into account greenhouse gas emissions and their impact on the global climate. These observations suggest that regulatory interventions are needed to induce new technologies that are not only more productive but also more sustainable (Box 6.4).

---

**Box 6.4   Induced Sustainable Innovation: The Case of Water-Saving Technologies**

Both economic and environmental regulations can raise the cost of pollution or the cost of overexploiting common property resources, and thus induce the development and adoption of more sustainable technologies. In many arid parts of the world, the use of diesel and electric pumps has resulted in water extraction exceeding recharge rates, causing the depth to groundwater to increase. This can raise the cost of irrigation, and in some cases, threaten the economic viability of agriculture, as well as cause land subsidence, dewatering of springs and wells, streamflow depletion, and seawater intrusion in coastal areas. For example, in Harney County, a high desert region of Central Oregon, from 2000 to 2015, the number of irrigated acres of pasture and hay production doubled, and the depth to groundwater increased by almost 100 feet. This is a classic example of the overuse of a common property resource, as there are many owners or users of land over the aquifer. Similarly, in most parts of the world, property rights to groundwater have not been established due, in part, to the difficulty of knowing how much water there is underground. As a result of the groundwater depletion in Harney County, the Oregon state government imposed a moratorium on issuance of additional permits for water pumping until the problem could be evaluated and solutions developed. A number of solutions have been proposed, including limiting the amount of water to be pumped by each permit holder, or creating a system of 'tradable' water allowances that would limit the total amount extracted, but would allow those producers wanting to increase their water use to buy allowances from others. Another solution is technological, notably the use of 'low-elevation sprinklers' that increase irrigation efficiency. This example illustrates how economic factors, such as the increased cost of pumping, as well as regulation of environmental externalities, for example, by restricting water use or by 'pricing' water through tradable allowances, can create incentives for development and adoption of more sustainable technologies.

## 6.4.2  Demand-Side Drivers of Sustainable Development

We argued in Sect. 6.2 that collective action is likely to be necessary to move an economy and its agricultural sector toward a more sustainable development pathway. Nevertheless, there is an important role for individual choices to play in moving toward sustainable development in market economies, by acting on the demand side of food markets and the signals that this can convey through market prices to farmers. In a nutshell, as general awareness of sustainable development goals increases, there should be a demand by consumers for food products produced using sustainable methods. If sustainably produced products are no more costly than ones produced with unsustainable methods, then markets solve the sustainability challenge by incentivizing firms to produce the more sustainable products that consumers want. But if this were true for most foods, we would automatically be on a sustainable agricultural development pathway and there wouldn't be a problem! So we have to conclude that when more and less sustainable products are competing in markets—say, a food produced with and without pesticides—the sustainable products will be supplied if consumers are willing to pay the higher prices for them that producers must charge to cover the higher cost. And if this is true, then over time, consumer demand for more sustainable products will create an incentive through the supply chain from retail to wholesale to farmers for more sustainable technologies to be developed and used.

So now let us see how this logic relates to the technologies we now have. The agricultural revolution of the twentieth century made possible an increase in the per capita availability of food calories throughout most of the world, and a decline in the real cost of those calories for consumers around the world. The main objective of these technological innovations was to increase the quantity of agricultural commodities. A food system coevolved with the industrialization of agriculture that transported, processed, and made available a wide array of food products that much of the world finds today in grocery stores. As we noted in Chap. 2, one of the important economic consequences of modern economic growth has been a rapid increase in per capita incomes, especially in the latter half of the twentieth century and continuing today, particularly in the transitional economies such as Brazil, China, and India. One of the most significant secondary effects of this increase in incomes is an increase in the value of people's time and the effects that has on their consumption behavior. This change explains the growth in the demand for an array of increasingly specialized goods and services, including ones that substitute for time spent preparing foods from raw ingredients in the home. So it is clear that economic signals from the demand side do indeed influence the food system throughout the supply chain from retail to wholesale. We can conclude that if the system were to send strong market signals of a demand for more sustainably produced foods, the system would convey that message to farmers and to the research and development system supporting agriculture. Indeed, the transition to a service- and information-based postindustrial economy is also associated with a rapidly declining cost of information that has profound impacts on the structure and functioning of markets at local, national, and global scales. These changes are also strongly associated with the globalization of the food system and the trends in agricultural development.

*Markets for Quality-Differentiated Products* Antle (1999) argues that these trends driven by economic development are leading to a reorientation of the demand side of the food systems around the world, from a historical focus on food *quantity* to an increasing focus on various *quality* attributes of food products. Among these quality attributes are not only the conventional taste, nutritional, and safety attributes, and the amount of time required for preparation, but also characteristics of the production process used such as the use of conventional or organic methods, the use of genetically modified crops or animals, and the way that animals are treated during the production process.

In other words, there appears to be a trend toward food being viewed by consumers as *quality-differentiated* products. How do these markets work? First, it is generally true that higher-quality products cost more to produce and so firms generally charge higher prices for them and these prices convey information about quality (otherwise, lower-quality products would not be produced). However, if information about product quality is not essentially free, or if markets are not highly competitive, firms may set prices above the cost of production and thus will not necessarily signal quality.

In some cases, product quality information is low cost; for example, a consumer can often tell the quality of fresh fruits and vegetables by visual inspection. Other attributes can only be discerned through use (sometimes called 'experience goods'); when buyers purchase the same product repeatedly, sellers can establish a reputation for quality. However, some quality attributes cannot be discerned from inspection or use. For example, consumers can't test foods for chemical or pathogen contamination. Even with pathogens, people often have difficulty knowing the cause of a health effect (e.g., gastrointestinal problems can be associated with any one of many foods consumed; cancer and other chronic effects are not realized until long after consumption). Such 'credence goods' require some other form of quality mechanism, such as product quality testing and certification by an independent entity. This means there needs to be a 'market for product quality,' and raises the question of how such markets work. Similarly, product quality attributes that are extrinsic to the product quality but instead are related to the process used, such as how animals are treated, or whether organic production methods are used, require some type of credible product quality information such as labeling or certification.

*Information-Based Policies.* The demand for quality-differentiated products creates a demand for product quality information. But is there such a market? Today consumers can obtain such information from a wide array of sources, some credible and some not, suggesting that there is a problem with information markets. The problem is that information is often a 'public good'—meaning that it can be used by one person without diminishing its value to other people (its use is 'non-rival'), and it is freely available. For example, information now available on the Internet about products and their attributes is largely a public good, but some of that information is not credible because it is obtained from the seller or an agent of the seller. So we can argue that *credible* information may be more appropriately defined as what is called a 'club good'; it is non-rival in use but it is not freely available. To the extent that such information is valued by consumers, they should be willing to pay for it, and markets for such information should arise where the value of information to consumers meets or exceeds the cost of producing it.

The fact that well-functioning markets for product quality information do not work for many kinds of products leads to a rationale for government intervention. For example, for foods with credence attributes such as safety, particularly cases where serious illness or death can occur as with fresh meats and seafood, safety regulations are often imposed by governments. These regulations often include process-based requirements, such as restrictions on pesticide applications before fruit and vegetable harvesting, the use of 'hazard analysis critical control point' systems for food processing, and product quality performance standards for automobiles.

Nevertheless, there is a strong economic rationale for governments to support product quality information markets to the extent feasible, for example, by supporting the creation of certification standards and product labeling. One important reason is that information-based policies subject the provision of quality attributes to the 'market test': when quality-differentiated products are available, people are free to choose between alternatives. Obviously, this is not relevant to cases where health and safety of the consumer are at stake. But when extrinsic attributes of the product are involved, such as the characteristics of the production process that do not directly affect the product's intrinsic quality, this seems appropriate. For example, many consumers want to buy products that are produced 'sustainably.' With a credible standard for product quality and certification, private firms can implement procedures to ensure production processes satisfy sustainability criteria, and the prices of certified products can incorporate the cost of complying with the standards and providing the certification. Many examples of certification are familiar to anyone visiting food stores today, including for organic and sustainably produced foods, fair trade products, and so on. This demand for product quality attributes is being supported now by new supply chain technologies, such as 'block chain' software, that improve the capability and lower the cost of tracking product quality attributes through supply chains from producers to consumers.

### 6.4.3  Supply-Side Policies and Sustainable Development

Evidence suggests that consumer preference for sustainably produced food can influence the agricultural development pathway and move it in a sustainable direction, particularly for production of food products that are closely connected to the consumer. For example, we see the market for sustainably produced beverages such as coffee and wine having an impact on some segments of the industry. Yet, in Sect. 6.2, we argued that this demand-side effect may not be sufficient.

Clearly, the sustainability challenges faced by the very poor, small, semi-subsistence farms described in Chap. 4, which have to do largely with their small size, poor soils, and lack of access to technology and markets, cannot be solved by participating in supply chains for quality-differentiated food products. Indeed, as we discussed in Chap. 3, there is a natural economic efficiency bias in commercial supply chains, from producers to retailers, toward larger-scale operations.

Likewise, there are good reasons why consumer demand alone may not be sufficient to address many of the sustainability challenges associated with large-scale, commercial food commodity production discussed in Chap. 5. One reason is that consumers may be free riders—they may not be willing to pay more for sustainably produced products if the sustainability outcomes they would be paying for can be achieved by other peoples' actions. Another reason is that consumers may not be well informed or may not understand some aspects of sustainability, such as the value of biodiversity or limiting climate change.

There are similar incentive problems on the supply side, as our earlier discussion of fertilizer use and water pollution illustrates. The issue is not that farmers don't care about pollution; rather the issue is that they would have to bear a cost to reduce pollution, and that they perceive that their individual actions would not have any discernible effect on the problem. Thus, in many cases they choose—one could say, from their perspective, rationally—not to voluntarily reduce their use of fertilizer.

A number of policy instruments have been developed to deal with environmental externalities. The most obvious solution is a legalistic one: if pollution is a problem, why not just make it illegal? In some cases, that is a feasible and even best approach. For example, allowing almost any amount of highly toxic substances such as arsenic to be intentionally emitted into water bodies is a crime in most countries with severe penalties. But in many cases, such as air pollution by industrial process, different plants emit different amounts of pollution, and 'zero' pollution is not a feasible option. Imposing a uniform emission limit on each plant is an obvious solution, and this is the solution that was used in early environmental regulations. However, as we discuss in Sect. 6.3, imposing uniform pollution limits on each polluter is often more costly than necessary to achieve the desired reduction in pollution. This is particularly true for some types of agricultural pollution, such as water pollution by nutrients or pesticides, because of the highly heterogeneous conditions of farms (different soils, proximity to water, and so on).

One of the solutions proposed by economists to address externality problems such as pollution is to use economic incentives to encourage individual consumers and firms to 'internalize' externalities—that is, to take into account the social costs of the externality in their consumption or production choices. These incentives can take a number of forms. Many people argue that a 'polluter pays' principle should be followed in designing such incentives. This could be done by taxing sellers of polluting products, or in the case of a polluting input such as fertilizers or pesticides, taxing the input. A widely discussed approach to climate policy is to create a 'carbon tax' based on the carbon dioxide emissions associated with fossil fuels, for example. The tax approach makes sense for greenhouse gas emissions because regardless of where the emissions occur, they mix in the atmosphere and have essentially the same impact on the global climate. However, in other cases, this is not true. Notably, in the case of agriculture, a uniform tax does not account for the widely varying damages to water quality caused by the use of inputs such as fertilizers or pesticides, the damage that soil erosion has on water quality, or the impact that odors from livestock feedlots or pig farms have on surrounding communities. Ideally, the tax would be related to the harm caused by the pollution. But in agriculture, this is often impractical because agricultural pollution is 'non-point source,' meaning that it comes from many dispersed places, not from a 'point source' like a smoke stack or pipe discharging contaminants into a river, or its impacts vary, as in the case of the odors emanating from livestock operations.

Despite the seeming equity in making a polluter pay for the cost of pollution, another important lesson from economics is that the same outcome can be achieved by paying a polluter to reduce pollution. For example, in some agricultural regions, 'right to farm' laws are passed that give farmers the 'right' to use fertilizers and pesticides, make noise, and emit odors. In these situations, farmers can be induced to change their behavior, for example, to use less fertilizer, by offering a payment or subsidy to compensate for the economic loss they would bear from lower crop yields and lower profits. This approach is very similar to what is often called 'payments for ecosystem services.' (Schomers and Matzdorf 2013). We think of farmers as providing ecosystem services in the form of food production and land management that provide services such as clean water and wildlife habitat. To the extent that there are tradeoffs among these services, it makes sense to compensate farmers for providing more of one service that other people in society value—such as clean water—in exchange for less of another service such as food. One of the well-known examples of this approach is New York City's policy of compensating farmers in the watershed to reduce water pollution where the city obtains its drinking water. This approach was found to be much less expensive than filtering the polluted water. So even though some people may not think it is 'fair' to pay farmers not to pollute, it may be the most cost-effective way to achieve the desired outcome.

One way to efficiently implement this type of payment scheme is to have farmers submit 'bids' to participate. This has been done successfully in the Conservation Reserve Program in the United States. A challenge in constructing these kinds of incentive schemes is that agricultural activities often have multiple environmental dimensions. For example, placing highly erodible land into a conserving use, as in the Conservation Reserve Program in the United States, not only reduces soil erosion but also water contamination from fertilizers and pesticides, greenhouse gas emissions, and creates wildlife habitat. The approach taken by the US government is to create an 'environmental benefits index' that is used to identify land to be included in the program.

Other ways to limit externalities are to assign property rights or create liability laws that allow those who are harmed by pollution to sue the polluters for compensation. In agriculture, this is also often not feasible because of the 'non-point' nature of agricultural pollution—it is very difficult to attribute groundwater contamination to any one farmer, for example. Another solution is for communities to self-organize to find solutions, for example, to allocate rights to common property resources so that they are not overutilized. This is the solution studied by Nobel Laureate Elinor Ostrom. The limitation to this solution is that it works mainly in situations with relatively small numbers of individuals; otherwise, the 'transaction costs' associated with negotiating and enforcing agreements become prohibitive.

A final but important point is that all of these policies have both short-run and long-run impacts. At the farm level, policies that raise the cost of polluting activities or reward beneficial practices encourage changes in management 'on the intensive margin,' such as how much fertilizer is applied to a crop, as well as on the 'extensive margin' in terms of what crops are grown and the kinds of capital investments that are made in physical and human capita. Also importantly, these policies induce the public and private entities that develop new technologies to invest in ones that help reduce negative externalities such as pollution and increase positive ecosystem services. This induced innovation process plays a key role in the movement of agriculture toward more sustainable development pathways.

**Box 6.5    Toward Sustainable Dryland Wheat Systems in a 1.5 °C World**

Can the dryland wheat systems of the US Pacific Northwest (Box 3.1) be moved in more sustainable directions through the adoption of the currently available technologies of no-till cultivation of wheat, incorporation of a biofuel crop into the system, and payments for soil carbon sequestration? Analysis of these options under alternative future projections of climate, crop prices, biofuel prices, and payments for carbon sequestration suggests that this is both economically and environmentally feasible—but also illustrates the challenges in the policy dimension.

Analysis of this question was carried out for the 2020–2030 time horizon, assuming: wheat prices in the range of 25% below or above recent historical prices; biofuel prices ranging from 30% below to 100% above recent prices; and carbon prices ranging from 50% below or above the estimated 'social cost of carbon' of US$50 per metric ton carbon dioxide ($CO_2$) (Antle et al. 2018). Table 6.2 presents results for the lowest- and highest-price scenarios. For low prices, less than 50% of farms would be willing to adopt this system, assuming farms use the most profitable system. Those farms that benefitted from adoption would earn substantially higher incomes and also reduce the global warming potential of the system by 17–20%. However, in the high-price scenario, more than 70% of farms would be willing to adopt based on profitability, and global warming potential of the system would be reduced by about one-third. Also, environmental benefits would result from reduced soil erosion.

This example illustrates that improvements can be made in the sustainability of large-scale, commercial cropping systems. But it also raises the question of whether such changes would be feasible unless the political climate were to be more favorable than it has been in recent experience in the United States which lacks policies favoring climate mitigation. This research indicates that without the political will to substantially subsidize this type of system with high biofuel and carbon prices, many farmers would not be willing to adopt.

**Table 6.2** Adoption and sustainability impact of no-till wheat-camelina-fallow system in US Pacific Northwest dryland wheat region

| Scenario | Farm Size | Adopters (%) | Adopter Impact on Farm Income (%) | Change in Soil Emissions (%) | Change in System GWP ($) |
|----------|-----------|--------------|-----------------------------------|------------------------------|--------------------------|
| Low prices | Large | 41.623 | 208.976 | −73.739 | −17.647 |
|  | Small | 45.671 | 387.433 | −81.114 | −19.439 |
| High Prices | Large | 72.302 | 303.789 | −131.171 | −34.721 |
|  | Small | 74.266 | 577.479 | −134.888 | −35.816 |

Source: Based on data from Antle et al. (2018)
Note: GWP = global warming potential

**Box 6.6 Pathways to Sustainable Development in Mixed Crop-Livestock Systems of Kenya**

Simulation models were used to evaluate the potential for the crop-livestock systems in the Machakos region of Kenya (Box 3.2) to move toward a sustainable trajectory with lower poverty while also increasing and stabilizing soil productivity, assuming the Kenyan government implemented policies in its Vision 2030 plan for economic development. Two 'Representative Agricultural Pathways' (RAPs) were constructed for future policies, one less optimistic (RAP 1) and one more optimistic (RAP 2) in terms of improvements in market access and technology investments (Box 6.3). Figure 6.3 shows 'tradeoff curves' between the number of farms above the poverty line and changes in soil nutrients, as maize prices are varied from low to high values, for the entire Machakos region and for a zone with irrigated vegetable production. The current system (BAU) is not sustainable, as there is high poverty and ongoing nutrient losses. There is also a strong tradeoff between these two dimensions of sustainability, as higher maize prices raise incomes and increase the number of farms above the poverty line but result in even higher nutrient losses. This occurs because under BAU conditions, higher maize prices encourage maize production to increase without the additional fertilizer needed to maintain soil productivity. The BAU system also shows large differences in poverty between the entire region and the more productive irrigated zone. The Vision 2030 policy and technology interventions (RAPs 1 and 2) which increase fertilizer availability and lower fertilizer prices have a dramatic impact on the sustainability of the system. They shift the tradeoff curves in a more sustainable direction (move the curves outward), and also shift the relationship between income and soil nutrients to a positive 'win-win' or synergy relationship, indicating that with higher maize prices, farmers expand maize production sustainably by applying enough nutrients to avoid soil degradation. This is true for the entire region as well as for the irrigated system.

*(continued)*

**Box 6.6** (continued)

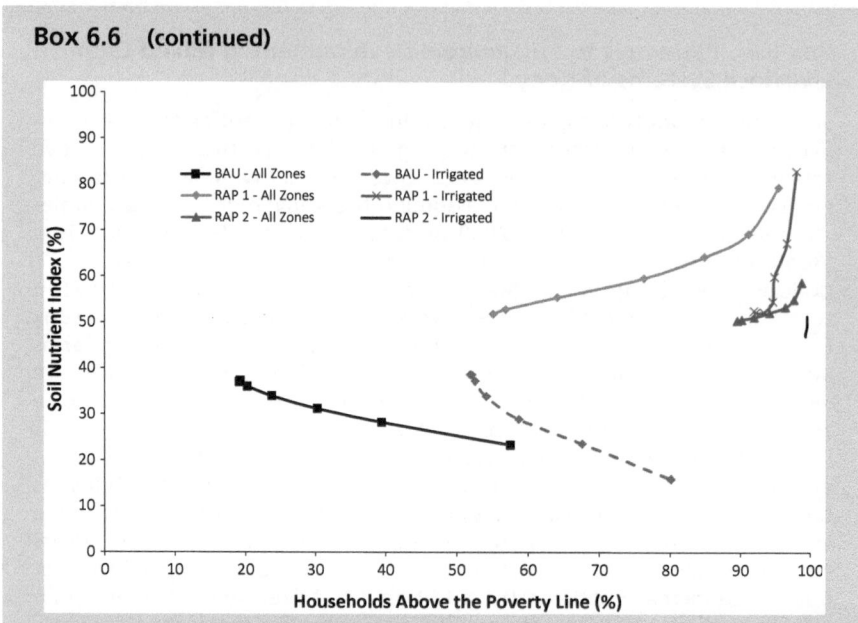

**Fig. 6.3** Poverty-nutrient tradeoffs and synergies in Machakos, Kenya for BAU and Vision 2030 Scenarios. (Note: BAU = business as usual; RAP 1 = low policy effectiveness; RAP 2 = high policy effectiveness. Tradeoffs generated by changes in the mean maize price (maize price increases from left to right, mid-point is base price). Soil Nutrient Index less than fifty implies nutrient loss, greater than fifty implies nutrient gain. Source: Based on Valdivia et al. (2017))

## 6.5 Conclusion

In this chapter, we discuss the ways that we can, individually and collectively, move agricultural development in more sustainable directions. This complexity means that simple, one-size-fits-all solutions to sustainable development are unlikely, so solutions must be tailored to fit the specific system, location, and problem scale. We discuss the two essential steps in moving the agricultural sector and agricultural systems toward more sustainable pathways. First is a design step that involves participatory processes to select indicators and set goals, and also involves the evaluation of pathways toward those goals using data and modeling tools. Second is an implementation step that involves incentivizing more sustainable technologies, as well as demand-side and supply-side policies that incentivize changes in the behavior of consumers and producers. We illustrate the challenges to sustainable development with examples discussed in previous chapters.

We conclude on a cautiously optimistic note. The tools needed to design and implement more sustainable agricultural development pathways for both developing and industrialized countries are advancing rapidly and becoming widely available. However, it is also clear that there are great political and governance challenges to be overcome at global, national, and local levels for both the design and implementation of more sustainable development pathways. It is clear that agriculture and the food system must change—the case of deforestation we highlighted in Chap. 5 being one of the many examples we discuss in this book. Recent studies show that substantially reducing meat consumption could have a large impact on greenhouse gas emissions and have other environmental and health benefits (EAT-Lancet Commission 2019). But meat production, consumption, and trade are growing rapidly, particularly in the regions of the world undergoing the transition from low to middle-income status, and many people argue that increasing meat consumption can play a valuable role in improving the nutrition in poor regions of the world. What is lacking in the debate over sustainable development, in our view, is to go beyond the identification of goals and the design of possible sustainable development pathways, to the implementation of feasible actions—technologies and policies—that will move today's agricultural systems in more sustainable directions. We have the tools, now we must use them.

# References

Antle, J.M. 1999. The New Economics of Agriculture. *American Journal of Agricultural Economics* 81 (5): 993–1010.

Antle, J.M., S. Cho, H. Tabatabaie, and R. Valdivia. 2018. Economic and Environmental Performance of Dryland Wheat Systems in a 1.5 Degree C

World. *Mitigation and Adaptation Strategies for Global Change* 24: 165–180. https://doi.org/10.1007/s11027-018-9804-1.

Antle, J., J. Jones, and C. Rosenzweig. 2017. Next Generation Agricultural System Data, Models and Knowledge Products: Synthesis and Strategy. *Agricultural Systems* 155: 179–185.

Antle, J.M., and C.O. Stöckle. 2017. Climate Impacts on Agriculture: Insights from Agronomic-Economic Analysis. *Review of Environmental Economics and Policy* 11 (2): 299–318. https://doi.org/10.1093/reep/rex012.

EAT-Lancet Commission. 2019. Summary Report of the EAT-Lancet Commission. https://eatforum.org/content/uploads/2019/07/EAT-Lancet_Commission_Summary_Report.pdf.

FABLE. 2019. *Pathways to Sustainable Land-Use and Food Systems. 2019 Report of the FABLE Consortium*. Laxenburg and Paris: International Institute for Applied Systems Analysis (IIASA) and Sustainable Development Solutions Network (SDSN).

Hayami, Y., and V.W. Ruttan. 1971. *Agricultural Development: An International Perspective* (2nd ed., 1985). Baltimore: The Johns Hopkins Press.

Heady, E.O. 1961. *Agricultural Production Functions*. Ames, IA: Iowa State University Press.

Johansson, Robert, Mark Peters, and Robert House. 2007. Regional Environment and Agriculture Programming Model. TB-1916. U.S. Department of Agriculture, Economic Research Service.

Jones, J.W., J.M. Antle, B.O. Basso, K. Boote, R.T. Conant, I. Foster, H.C.J. Godfray, M. Herrero, R.E. Howitt, S. Janssen, B.A. Keating, R. Munoz-Carpena, C. Porter, C.E. Rosenzweig, and T.R. Wheeler. 2017. Towards a New Generation of Agricultural System Models, Data, and Knowledge Products: State of Agricultural Systems Science. *Agricultural Systems* 155: 268–288.

Nelson, G.C., H. Valin, R.D. Sands, P. Havlik, H. Ahammad, D. Deryng, J. Elliott, et al. 2013. Climate Change Effects on Agriculture: Economic Responses to Biophysical Shocks. *Proceedings of the National Academy of Sciences of the United States of America* 111: 3274–3279. https://doi.org/10.1073/pnas.1222465110.

Rosenzweig, C., J. Antle, and J. Elliott. 2016. Assessing Impacts of Climate Change on Food Security Worldwide. *Eos* 97. https://doi.org/10.1029/2016EO047387. Published on 9 March 2016.

Rosenzweig, C., J.W. Jones, J.L. Hatfield, A.C. Ruane, K.J. Boote, P. Thorburn, J.M. Antle, et al. 2013. The Agricultural Model Intercomparison and

Improvement Project (AgMIP): Protocols and Pilot Studies. *Agricultural and Forest Meteorology* 170: 166–182.

Rosenzweig, C., A.C. Ruane, J.M. Antle, et al. 2018. Coordinating AgMIP Data and Models Across Global and Regional Scales for 1.5° and 2.0° C Assessments. *Philosophical Transactions of the Royal Society A* 376: 20160455. https://doi.org/10.1098/rsta.2016b.0455.

Schomers, Sarah, and Bettina Matzdorf. 2013. Payments for Ecosystem Services: A Review and Comparison of Developing and Industrialized Countries. *Ecosystem Services* 6: 16–30.

Valdivia, R.O., J.M. Antle, C. Rosenzweig, A.C. Ruane, J. Vervoort, M. Ashfaq, I. Hathie, S. Homann-Kee Tui, R. Mulwa, C. Nhemachena, P. Ponnusamy, H. Rasnayaka, and H. Singh. 2015. Representative Agricultural Pathways and Scenarios for Regional Integrated Assessment of Climate Change Impact, Vulnerability and Adaptation. In *Handbook of Climate Change and Agroecosystems: The Agricultural Model Intercomparison and Improvement Project Integrated Crop and Economic Assessments, Part 1*, ed. C. Rosenzweig and D. Hillel. London: Imperial College Press.

Valdivia, R.O., J.M. Antle, and J.J. Stoorvogel. 2017. Designing and Evaluating Sustainable Development Pathways for Semi-Subsistence Crop-Livestock Systems: Lessons from Kenya. *Agricultural Economics* 48 (S1): 11–26. https://doi.org/10.1111/agec.12383.

van Ittersum, Martin K., Frank Ewert, Thomas Heckelei, Jacques Wery, Johanna Alkan Olsson, Erling Andersen, Irina Bezlepkina, Floor Brouwer, Marcello Donatelli, Guillermo Flichman, Lennart Olsson, Andrea E. Rizzoli, Tamme van der Wal, Jan Erik Wien, and Joost Wolf. 2008. Integrated Assessment of Agricultural Systems—A Component-based Framework for the European Union (SEAMLESS). *Agricultural Systems* 96 (1–3): 150–165.

van Wijk, M., M. Rufino, D. Enahoro, D. Parsons, S. Silvestri, R. Valdivia, and M. Herrero. 2014. Farm Household Models to Analyse Food Security in a Changing Climate: A Review. *Global Food Security* 3: 77–84.

World Bank. 2019. Intended Nationally Determined Contributions (INDCs). http://spappssecext.worldbank.org/sites/indc/Pages/INDCHome.aspx.

# Index